Great
Philosophical
Objections to
Artificial
Intelligence

Great Philosophical Objections to Artificial Intelligence

The History and Legacy of the AI Wars

Eric Dietrich, Chris Fields, John P. Sullins, Bram van Heuveln and Robin Zebrowski

BLOOMSBURY ACADEMIC
LONDON · NEW YORK · OXFORD · NEW DELHI · SYDNEY

BLOOMSBURY ACADEMIC
Bloomsbury Publishing Plc
50 Bedford Square, London, WC1B 3DP, UK
1385 Broadway, New York, NY 10018, USA

BLOOMSBURY, BLOOMSBURY ACADEMIC and the Diana logo are trademarks of
Bloomsbury Publishing Plc

First published in Great Britain 2021

Cover design by Toby Way (tobyway.co.uk)
Cover image © Getty Images / Bletchley Park Trust / Contributor

A catalogue record for this book is available from the British Library.

Library of Congress Cataloging-in-Publication Data
Names: Dietrich, Eric, author. | Fields, Chris, author. | Sullins, John P., author. |
Van Heuveln, Bram, author. | Zebrowski, Robin, author.
Title: Great philosophical objections to artificial intelligence : the history and
legacy of the AI wars / Eric Dietrich, Chris Fields, John P. Sullins,
Bram Van Heuveln and Robin Zebrowski.
Description: London ; New York : Bloomsbury Academic, 2021. |
Includes bibliographical references and index.
Identifiers: LCCN 2020036916 (print) | LCCN 2020036917 (ebook) |
ISBN 9781474257114 (hardback) | ISBN 9781474257107 (paperback) |
ISBN 9781474257091 (epub) | ISBN 9781474257077 (ebook)
Subjects: LCSH: Artificial intelligence–Philosophy.
Classification: LCC Q335 .D54 2021 (print) | LCC Q335 (ebook) | DDC 006.301—dc23
LC record available at https://lccn.loc.gov/2020036916
LC ebook record available at https://lccn.loc.gov/2020036917

ISBN: HB: 978-1-4742-5711-4
       PB: 978-1-4742-5710-7
       ePDF: 978-1-4742-5707-7
       eBook: 978-1-4742-5709-1

Typeset by RefineCatch Limited, Bungay, Suffolk
Printed and bound in Great Britain

To find out more about our authors and books visit www.bloomsbury.com
and sign up for our newsletters.

Contents

List of Figures

Prologue: The AI Wars and Beyond

In 1978, the philosopher and AI researcher, Aaron Sloman, wrote:

> I am prepared to go so far as to say that within a few years, if there remain any philosophers who are not familiar with some of the main developments in artificial intelligence, it will be fair to accuse them of professional incompetence, and that to teach courses in philosophy of mind, epistemology, aesthetics, philosophy of science, philosophy of language, ethics, metaphysics, and other main areas of philosophy, without discussing the relevant aspects of artificial intelligence will be as irresponsible as giving a degree course in physics which includes no quantum theory (section 1.2, p. 3).
>
> Aaron Sloman, 1978, *The Computer Revolution in Philosophy: Philosophy of Science and Models of the Mind*

Sloman's prediction failed … spectacularly. Here in the early part of the twenty-first century, many of today's most well-known and distinguished philosophers are happily unfamiliar with any technical developments in AI. And many philosophy courses of all types, including philosophy of mind, are taught today with only a passing reference to artificial intelligence. In fact, it is fair to say that *most* philosophy courses never mention any AI. Where AI is mentioned, moreover, the issues of primary interest, e.g. the ethics of job replacement or of robot caregivers in nursing homes, are issues that were of minor interest, at best, in the late 1970s. What happened? How could such an explicit and robust prediction about such a promising technology and science be so wrong?

Of course, the well-known phenomenon of early, over-zealous enthusiasm for a new discovery or unproven technology is a partial explanation. But a deeper explanation is that what would have been the greatest engineering triumph in all of human history – the invention of machines with human-level intelligence – fizzled. While AI often makes the news, it is not because human-level AI has been achieved. Indeed, nothing like human-level AI exists. The story of how and why the first wave of AI fizzled, why philosophers worked so hard in the service of the fizzling, and how a different sort of second or even third wave of AI is nevertheless flourishing today is told here.

The AI wars of the late twentieth century, which philosophers started, also fizzled: they were fought to a draw. Neither the pro-AI side nor the anti-AI side won any significant victories over the course of the five decades the wars were fought. While the issues that the wars were about are still debated, neither the pro- nor the anti-AI arguments have changed significantly since they were first introduced. Both sides, moreover, now take contradictory credit for the state of AI today. 'Human-level AI has failed', assert (retired) naysaying philosophers, 'and we know why'. 'AI is stronger than ever and will soon produce a computer or robot with human-level intelligence', say the (descendants of the first) AI researchers. We sort out this odd situation in Part I of this book, using the happy fact that the fizzling of first-wave AI and the fizzling of the philosophical AI wars are closely related. We then turn, in Part II, to an examination of the major philosophical issues that have arisen since the AI wars, which incorporate some of the concerns and arguments that drove the original debates but place them in an entirely new setting.

Between roughly 1950 and 2000, there were four central AI wars:

1　The war over the logical limitations of computers, which were thought to prevent AI completely;
2　The war over which architecture would allow us to implement machine intelligence, assuming we could;
3　The war over whether computers could ever think *about* anything at all – mousetraps don't think about mice, and computers are just fancy mousetraps, aren't they?
4　The war over whether computers could be creative and figure out any of the subtle relevancy connections that lace our world together.

We review these in detail in Part I. We discuss not only how the wars were fought, but how they ended, and why they ended in stalemate, not in victory for one side or the other nor in consensus.

Beyond the AI wars, there is an ongoing story that affects all of us every day. If you have ever used a search engine to search the web, or just visited a webpage or used a smart phone, you have interacted with an artificial intelligence. Was AI lurking behind the scenes? Yes. Does this AI have (or could it have) nefarious intentions of its own (beyond those perhaps held by the humans running the corporations that deploy the most AI)? This is an unresolved issue as of the early part of the twenty-first century. It challenges not only our ethics as users and potential builders of AIs, but also our ideas about what kinds of systems could *have* ethics, whether benign or nefarious.

Interestingly, but in retrospect, not surprisingly at all, the idea that a computer could be as intelligent as a human, that a computer could have a mind, and that a computer could be conscious and act as an ethical agent were proposed almost as soon as general-purpose computers came into existence. This was a natural, even inevitable, proposal given what general-purpose computers could do and could be envisioned doing. The AI wars of the late twentieth century were concerned mainly with the first of these ideas: the idea that computers could be intelligent. While this question was never resolved, it is no longer of primary interest. Questions about AI consciousness and ethics have taken its place. We consider in Part II how these questions have come to the fore, and why.

Part I

The AI Wars, 1950 to 2000

Introduction

The first AI conference was held in 1956 at Dartmouth College, and one of the first AI programs, Logic Theorist, was completed that same year. As with all such conferences about a new development, the conference happened years after work on artificial intelligence had started – though before 1956, the field was not called 'artificial intelligence' nor was it even regarded as a separate field. However, just a few years later, and continuing for decades, the AI project of building a machine with human-level intelligence was met with a barrage of sophisticated attacks by philosophers.

We can date the first of these attacks to around 1959, when J. R. Lucas, a British philosopher, presented his paper 'Minds, Machines, and Gödel' to the Oxford Philosophical Society.[1] Taken together, these attacks pointed out what appeared to be limitations to computation, raised problems about machines being able to think about specific things like coffee cups or numbers, and flagged the general problem of how a machine could be conscious and aware. Whether or not a computer could be rational and moral was also questioned. And finally, several issues were raised regarding cognitive architecture. For example, perhaps only something with an architecture like a brain could actually be intelligent, and computer architectures are nothing like brain architectures. Of course, AI researchers and pro-AI philosophers responded. Sometimes they responded by trying to directly refute the philosophical objections, other times they built and implemented computer programs.

We are concerned, in this book, more with the philosophical arguments than the implementations; the latter will be mentioned only when needed. We consider 'philosophy', moreover, to be an activity, not just an academic discipline. Many computer scientists and other AI researchers took straightforwardly philosophical positions and made philosophical arguments, often from a position of great naiveté about philosophy. Especially early in the wars, philosophers often responded in kind, from positions of great naiveté about computer science.

Somewhere around the turn of the millennium, the attacks on AI by philosophers abated; the pro-AI side also calmed down. No qualitatively new issues were identified and no new arguments were launched. While the old arguments were sometimes repeated, perhaps with small variations, they had largely lost their urgency. Did AI researchers successfully answer all the philosophers' objections and allay all their concerns? Far from it. Did AI researchers come to see that the philosophers were right? No. Did anti-AI philosophers come to see that they were wrong? No. Did AI researchers give up their quest? Not at all: research in AI techniques such as machine learning and data mining is robust and thriving, and its practitioners are currently very much in demand. Indeed it is mainly the success of AI in practice that has generated the ethical issues explored in Part II.

It is important to realize that not all philosophers were anti-AI. Many were very supportive and enthusiastic, like Sloman, above. These pro-AI philosophers, as well as AI researchers themselves, pushed back against the anti-AI philosophers. But, as mentioned above, no side succeeded in pushing the other off the field – genuine peace has not emerged. Rather a stalemate has arisen, along with the emergence of a wait-and-see attitude. As will be discussed in Part II, this attitude of wait and see was induced, at least in part, by the emergence on the scene of a major new player, cognitive neuroscience.

Here is a general overview of some of the different forces that together worked to quiet the AI wars.

1   In the beginning, there were many proclamations like Sloman's above. Many on the pro-AI side were positively gushing about how wonderful AI was and was going to be. The final hurdles to understanding human intelligence – a goal sought, arguably, since at least Plato – were falling . . . the end was in sight, true understanding was at hand. And with it, all the good things that would come from having artificial intelligences helping us run the world. The end of war (the bloody kind), the end of disease, famine, and hardship. However, the most important thing accomplished by the first wave of AI, actually, was teaching us how unbelievably complicated the hardware of the brain is, and how unbelievably complicated the processes involved in thinking really are. Basically, AI researchers and their comrades underestimated by several orders of magnitude how hard it was going to be to build a machine with human-level intelligence. As the decades rolled by, this failure of AI to deliver our intelligent, silicon planet-mates struck many anti-AI philosophers as evidence that they, the anti-AI-ers, were right or at least in the right neighbourhood.

2   As the overwhelming and completely under-appreciated complexity of human thought was emerging and making everyone re-evaluate positions once thought obvious, AI was quietly progressing. The successes of Google and Facebook, as corporations, are due to these advances. But this kind of progress didn't seem philosophically problematic – it wasn't a threat to our metaphysical or our epistemic

understandings of ourselves. AI researchers involved in this latter-day progress did not go around saying their machines were conscious, they didn't even say their machines were intelligent. They merely said their machines were more useful. What's philosophically objectionable about that? It was not until this kind of second-wave AI was socially ubiquitous that ethical issues about the uses of AI fully came to the fore.

3   We have so far referred to the sides in the AI wars as the pro-AI side and the anti-AI side. This is convenient, but unfortunately it gives the impression that each side was coherent and of one mind. This impression is wrong. There were rebellions and robust disagreements between members of the same sides. Philosophers who thought AI was a pipe dream for one reason attempted to refute those who thought it was a pipe dream for another reason. AI defenders defended different types of AI (e.g. classical, rule-based systems versus distributed, parallel systems), and many held quite different views of what would constitute an AI success. The topic of AI seemed to unleash a storm of arguments all going in myriad directions. All this tumult was exhausting.

4   The philosophical attacks not only exposed problems in AI and the related fields of cognitive and developmental psychology, and more recently cognitive neuroscience, but also exposed problems in philosophy. Key philosophical concepts having to do with creativity and rationality, with semantics and thinking about objects in the environment, with consciousness, and with the very notion of having a mind at all were deployed to attack AI. But then other philosophers argued that these very notions were themselves open to attack from different directions, the most important being that the major philosophical concepts were not well-defined. Soon it emerged that it may not be possible to define these concepts well enough to use them. It's hard to win a battle if you are unclear on how to use your weapons. It's even harder if you aren't sure what your weapons are.

There is much more to be said about these four. Fortunately, there is a book in which to say it all: this book. Understanding how these four forces played out in the decades leading up to 2000 is the goal of Part I.

The AI wars were a short, but heady time in human history. The details of why the wars went silent reveal a goldmine of information and knowledge about humans and their neuropsychology, AI and computers, philosophy and its strange nature, and the roles of science and technology in our modern culture. Here then is the history of the AI wars.

The First War: Is AI Even Possible?

Gödel and a Foundational Objection to AI

1. Introduction: Advances and Naysayers

Human technological advances have always come with naysayers opposing the advance. Sometimes the opposition raises good points. Splitting the atom was such a case. In hindsight, one can rationally conclude that the proliferation of nuclear weapons now was too high a price for nuclear knowledge then. Sometimes the naysayers, as well-meaning as they are, miss the profound problem for some immediate one, usually because they lack the knowledge future advances will bring. At the turn of the twentieth century, those opposed to the automobile decried the speed and danger of the machines. Speed and danger proved to be real problems, of course, but the profound problem was automobiles dumping the greenhouse gas carbon dioxide into our atmosphere. Today, the typical passenger vehicle dumps into our air around 4.6 metric tons of carbon dioxide per year.[2]

There are the puzzling cases. In one of the crueller ironies of history, the great Greek physician, Galen, became his own naysayer. Galen (c. 130–200 CE) wrote several important books on medicine, the most influential of which was called *On the Usefulness of the Parts of the Body*. Galen's influence was due partly to the fact that he was one of the first experimental physicians, and he constantly urged the physicians who came after him to learn from experience and to focus on knowledge that could cure patients. Unfortunately, Galen's influence went beyond anything he could have imagined, beyond anything he would have wanted. For approximately fifteen hundred years, until the late seventeenth century, Galen's books were regarded as sacred texts. Instead of furthering the field of anatomy by dissecting human bodies, physicians read Galen. For all those centuries, medicine was not a science, but a branch of philology. Anatomy and diagnosis were done by studying what Galen said, and when what he said didn't fit the facts, physicians redoubled their efforts to figure out what he really meant. Physicians ignored the data that were right in front of them: bodies.

Sometimes, the advance cuts so deep into our vision of ourselves and our humanity that opposition becomes mandatory and fierce. In 1633, the Catholic Church opposed Galileo's heliocentric theory of the solar system, as well as his use of his invention, the telescope, by threatening him with torture. And to this day, a majority of westerners reject Darwin's theory of evolution (Miller et al., 2006). Artificial intelligence is, unsurprisingly, another example of this sort of case. It hit some people with the force of Galileo's discoveries, with the horror of Darwin's discovery of evolution.

2. John Randolph Lucas: AI's First Naysayer

In 1956, John McCarthy first coined the term 'artificial intelligence'. Just three years later, in 1959, Oxford philosopher John Lucas presented a paper at the Oxford Philosophical Society arguing that for strictly logical reasons, AI was impossible. This was the first anti-AI salvo. Lucas said:

> Gödel's theorem seems to me to prove that Mechanism is false, that is, that minds cannot be explained as machines . . . Gödel's theorem states that in any

consistent system which is strong enough to produce simple arithmetic there are formulae which cannot be proved-in-the-system, but which we can see to be true.

1961, p. 112

Just as round squares are logically, hence *completely* impossible, so is computational psychology (basically, cognitive psychology), and hence, so is AI. If true, Lucas's conclusion would bring the entire pure AI project to a standstill (though applied AI, like we see today in marketing programs owned by Facebook, Google and Amazon, would still be with us). We now have to explain Gödel's theorem. Then we will return to Lucas's objection.

3. Gödel's Theorem

In 1930, the twenty-four-year-old Kurt Gödel presented his shocking result that there is no formal proof system that can prove all arithmetical truths. (Gödel presented this at the Second Conference on the Epistemology of the Exact Sciences, held in Königsberg, 5–7 September. His paper was published in 1931.) His result is known today as *Gödel's Incompleteness Theorem* (more technically, it is known as *Gödel's First Incompleteness Theorem*; see below). It is considered one of the most important mathematical results ever produced, as it was the first result to show an inherent limitation to the power of mathematics. Many regard Gödel's famous theorem as being as profound as Einstein's work on relativity.

Very briefly, and non-technically, Gödel developed a statement written in logic combined with the counting numbers and simple arithmetic (addition and multiplication) which stated (given here in English):

I am not provable.

Let's call this sentence G. And call Gödel's formal system of logic, numbers, and arithmetic A. Then Gödel pointed out that if we assume that A is *consistent* (i.e. that A is not self-contradictory – that it is not possible to prove some proposition P and its negation *Not-P*) then G follows immediately. For if it were possible to prove G, then A would be inconsistent: G says that it is not provable, but we would have just proven it. This is obviously contradictory. The assumption of consistency bars this, so G must not be provable. But this is what G says. Whence G is true . . . but not provable. Hence, A contains true statements that cannot be proved (this property is usually called *incompleteness*; for contrast, many logical, formal systems *without* numbers and arithmetic are *complete*, meaning that any true logical statement can be proved within the system; these systems are weaker than the one Gödel was working with).

The assumption of consistency is clearly doing quite a bit of work here. But it is an assumption that strikes mathematicians as so plausible that it is worth very little to question it. Never in the millennia-long history of arithmetic has anyone found any contradiction, nor even a hint of one. So, the consistency assumption is solid and well-accepted everywhere. But going further, Gödel's *Second Incompleteness Theorem* (which is really about consistency, but it is called the 'second incompleteness theorem') says that no formal system like *A* can prove its own consistency. So assuming consistency based on never ever seeing it violated is as good as we can do. But this is clearly good enough. No one doubts that *A* is consistent.

Lucas's Conclusion: *AI is impossible*. Since computers are formal systems, like Gödel's combination of logic, numbers, and arithmetic, there are statements we humans can prove but which a computer cannot.

4. What is Really Proved in Section 3.

Before proceeding, it is important to be clear about what we proved in section 3 (following Gödel). We proved this (in English):

If *A* is consistent then *G* is not provable.

That is, we have proved what logicians call a *conditional*, a statement of the form 'If X then Y.' One proves a conditional, any conditional, by assuming X and deriving Y. That is what we have done in section 3. By assuming *A*'s consistency, we can derive *G*'s unprovability. But we have to have *A*'s consistency to do that. Let's call the conditional here in section 4, G^*.

5. An Objection to Gödel's Incompleteness Theorem

We need to forestall one objection right away. In section 3, we proved (in a very nontechnical way) G^*, i.e. that *G* is true, but unprovable, given that *A* is consistent. Gödel proved his theorem technically and precisely. Nevertheless, we have proven G to be true even though it says it is not provable. So, we have proven A inconsistent. And so Gödel's theorem is false. So goes the objection.

No. The reason this objection is wrong is that neither we nor Gödel proved G^* within the formal system A. (This is a separate point from the one made in section 4.) That is, the proof of G^* is not itself in A, but in a *metasystem outside of A*. Gödel's proof is technically not a part of A; it is a part of *meta-logic* or *meta-arithmetic*. A is usually called the *object language*, while statements about A are made in a *meta-language*. The reason for this is that Gödel's proof is *about* arithmetical statements. Arithmetical statements, however, are about things in the real world, like numbers of crows or ravens, how many even prime numbers there are, etc. All natural languages, e.g. English and Swahili, collapse this distinction, so they make the distinction Gödel drew seem difficult. But it is not. For example, 'Dogs have 4 legs, usually' is a statement about dogs and legs. This is an arithmetical statement. But '"Dogs have 4 legs usually" has 5 words' is a meta-English, arithmetical statement about the sentence 'Dogs have 4 legs usually'. The reader probably considers both sentences to be sentences in English. And this is fine in our daily lives. But Gödel needed to use this distinction for his proof; Gödel is precisely using this distinction in his theorem. His proof is about arithmetical statements, not about dogs and the number of legs they have. This distinction will matter below.

6. Lucas' Objection Against AI

Lucas gave a sophisticated argument against the very notion of an intelligent machine based on Gödel's First Incompleteness Theorem:

> Gödel's theorem must apply to cybernetical machines [computers], because it is of the essence of being a machine, that it should be a concrete instantiation of a formal system. It follows that given any machine which is consistent and capable of doing simple arithmetic, there is a formula which it is incapable of producing as being true – i.e., the formula is unprovable-in-the-system – but which we can see to be true. It follows that no machine can be a [total] or adequate model of the mind, that minds are essentially different from machines.
>
> 1961, p. 112

Notice how this argument avoids making the false claim that there is one 'special' truth that cannot be proven by any formal system, but instead works with the correct claim that for every consistent formal system there is some sentence (at least one) that that system cannot prove. Indeed, the argument works by pointing out that given any specific system, one can always say: 'I cannot be *this* system, for *this* system cannot prove *that* formula ... but I can!' Since we can do this for any such system, we are therefore not any of them. And since any AI machine would have to be such a system (would have to be the physical implementation of such a system), the machine would be stuck with saying, 'I cannot prove this formula.'

7. Does Lucas's Argument Work?

The general consensus, though some disagree (see below), is that Lucas's argument does not work. Consider again G^*:

G^*: If A is consistent then G is not provable.

Lucas has given us an argument that, were a computer *bound* to its object language, then it could not prove G^*. We humans are not bound to our object language (at least we are not bound to A) so we can step out of A and prove G^*. However, Lucas has given us no argument that a computer cannot do just what we did, what Gödel did. Lucas has given us no reason to accept that a computer cannot abstract and move outside of object language A to a meta-language, the one in which G^* is couched. (We will see examples of computers abstracting in Chapter 3, when we examine learning in artificial neural nets.) Lucas merely assumes this limitation on computers. He is therefore simply begging the question here against AI – assuming what he wants to prove: he's assuming that machines cannot be as intelligent as humans. It is embarrassingly easy to prove what you want if you assume it is true.

But is there any reason to assume that computers cannot be programmed to step out of one object language to some broader meta-language? There appear to be none. None that aren't question begging, anyway.

So, it does seem as if computers could prove G^* just as we do, by stepping out of the relevant object language and into a new, higher meta-language. Stepping out of their object language does not prevent an AI from being a machine, contra Lucas.

8. Conclusion

Gödel proved a great and wonderful theorem. But Lucas, it seems, failed to show AI was impossible by using Gödel's theorem. And for principled reasons: there is no way to use Gödel's theorem to end AI. So, the project of building a computer with human-level intelligence can continue.

However, Lucas's quest is not dead. Roger Penrose, in two books, *The Emperor's New Mind* and *Shadows of the Mind* (1989, 1994), took up Lucas's idea, extended it significantly, and concluded that the human mind is not a computer because we have abilities computers cannot have. However, Penrose's arguments crucially involve *consciousness*, and, given that we currently have no idea what consciousness is, nor how any physical thing could be conscious, this makes evaluation of Penrose's arguments impossible (see Chapters 5 and 7 below, and Dietrich and Hardcastle, 2004). Computers may not be conscious now (is the internet conscious?), but no well-accepted reason has been given to believe they cannot be conscious. And for all we know, computers are conscious now. The one being used now to construct this chapter may be conscious even as this sentence is being typed.

2

How Would We Know If a Computer Was Intelligent?

The Turing Test is Not the Answer

Chapter Outline

1. Computing Machinery and Intelligence

Alan Turing started his famous and seminal paper 'Computing Machinery and Intelligence', which was published in 1950 in the philosophical journal *Mind,* with the following words (all unspecified quotes in this chapter are from this paper):

> I propose to consider the question, 'Can machines think?' This should begin with definitions of the meaning of the terms 'machine' and 'think.' The definitions might be framed so as to reflect so far as possible the normal use of the words, but this attitude is dangerous. If the meaning of the words 'machine' and 'think' are to be found by examining how they are commonly used, it is difficult to escape the conclusion that the meaning and the answer to the question, 'Can machines think?' is to be sought in a statistical survey such as a Gallup poll. But this is absurd. Instead of attempting such a definition I shall replace the question by another, which is closely related to it and is expressed in relatively unambiguous words.

It is considered by many to be the first serious intellectual analysis of the possibility of machine intelligence, and to this day remains a highly influential paper in philosophical discussions about the possibility of machine intelligence.

Turing's paper consists of two fairly distinct parts. In the first part, Turing describes what we nowadays call 'the Turing Test'. We provide a nonstandard, but we think correct, interpretation of this part. In the second part, Turing anticipates various objections to the possibility of machine intelligence, many of which are still commonly given nowadays. In this chapter, we will follow Turing's organization.

CHAPTER 2, DIVISION 1
THE TURING TEST

2. The Turing Test and the Imitation Game

Turing writes:

> The new form of the problem can be described in terms of a game which we call the imitation game.

Turing then describes the imitation game, which involves three people. A man and a woman are put in a room, while a third person, called the interrogator, of either gender, is put in a separate room. The interrogator is told that there are two people of different genders in the other room, and that the interrogator has to figure out which of the two people, known simply as A and B, is the man and which is the woman. Now, to make this an interesting game, the man (and let's suppose the man is A) is told to answer the questions as if he were a woman, while the woman is simply to answer truthfully. Thus, when asked 'Are you a woman?', both A and B will answer 'Yes'.

Turing then continues:

> We now ask the question, 'What will happen when a machine takes the part of A in this game?' Will the interrogator decide wrongly as often when the game is played like this as he does when the game is played between a man and a woman? These questions replace our original, 'Can machines think?'

The imitation game that Turing describes here quickly captured the imagination of the public. It was soon called Turing's Test, and is now popularly known as the Turing Test. The general idea is that a machine that does so well in the imitation game that the interrogator cannot distinguish (or at least has great trouble doing so) the machine's responses from those given by the human being, is said to pass the test.

The Turing Test has, over the decades, received mixed reactions. There have been many criticisms of the test, but there have also been defenders of the test. Before we enter any of the arguments for and against the test, though, we should get clear on exactly what it is that Turing is proposing. Or, more to the point: how should we think about the Turing Test?

3. What is Intelligence?

Some commentators on Turing's paper claim that with his imitation game, or Turing Test, Turing tried to create a working definition of intelligence. Indeed, fourteen years before this paper, Turing wrote another highly influential paper in which he defined what is now known as a Turing machine. This was a mathematical theory proposed to capture the intuitive notion of an *effective method*, or in other words, of a *computation*. Indeed, the machines that Turing talks about in the context of his imitation game are computers, rather than just any machine.

Here it suffices to point out that with his Turing machine, Turing gave a hard and precise definition for a fuzzy concept in the hopes of obtaining an objective and scientifically useful theory to explain and predict what could, and could not, be computed. In this, Turing succeeded spectacularly. His precise mathematical definition allowed him to prove important results in computability theory, and apparently his proposed definition (now known as Turing-computation) captured the notion of an effective method well enough that these mathematical results seem to describe the actual world we live in with a high degree of accuracy.

So, with his proposal of the imitation game, perhaps Turing was similarly hoping to obtain an objective way to classify a machine as belonging to an interesting class of objects: those that think and are intelligent. This would allow us to make useful scientific explanations and predictions regarding a machine's capabilities related to thinking and intelligence.

This is certainly an interesting way of looking at the Turing Test. In fact, the situation we find ourselves in can be compared to the case of the planets. People still (and maybe will forever) argue whether Pluto is a planet or not. However, by making a clear definition of what a planet is, we can at least make some objective and scientific headway in terms of making sense of the world we live in. Likewise, we can probably forever debate whether the Turing Test captures the notion of intelligence or thinking or not, but at least it seems to give us a practical and potentially more objective tool to make some important delineations about the objects in the world around us.

Moreover, by saying that Pluto is a dwarf-planet, and not a planet, we have started to make more fine-grained distinctions, ontologies, and taxonomies between the different objects in our solar system, helping us to obtain more precise and useful explanations and predictions. And Turing's notion of Turing-computation led to various other forms of computation, such as finite-state machines, and push-down automata, each with their own objective definitions and associated mathematical theorems, all reflective of what we hope are different natural kinds that exist in the world we actually live in.

Likewise, it is probably fruitful to differentiate between different kinds of intelligence. Indeed, by putting different constraints on how the imitation game is played, we might be able to differentiate between different kinds of intelligence. Was this what Turing proposed when he laid out the imitation game?

As exciting as the prospect sounds, this idea immediately goes against what Turing writes in the starting passage of his paper. Apparently, Turing does not think that we should try to answer the question 'Can machines think?' by trying to define such notions as thinking or intelligence. Indeed, he seems pretty explicit in his proposal to address any questions as to the possibility of machine intelligence by considering the imitation game *instead*. That is, Turing is looking for a way to address the question of machine intelligence that is able to *avoid* having to define intelligence, something that will provide us some guidance as to the question of whether or not machines can think, even as we don't know what thinking really *is*.

Of course, one could hold the view that instead of a theoretical definition, maybe Turing was still trying to provide some kind of *operational definition* of thinking and intelligence. For example, counting the number of cycles of radiation a cesium-133 atom produces doesn't tell us what time *is*, but nevertheless provides us with a measurement as to how much time has passed. Likewise, the imitation game does not provide us with a definition of what thinking and intelligence *are*, but still provides us with a practical way of determining whether something is intelligent or not. So, is *that* what Turing proposed?

One immediate thing to note about this interpretation of the Turing Test is that it cannot be an operational definition of intelligence, in general, since any entity (machine or not) whose intelligence is quite different from a human would never be able to pass the Turing Test. In fact, the Turing Test has sometimes been criticized for exactly this fact: the test is seen as too anthropocentric, since through its set-up it takes human intelligence as the standard of intelligence. On the other hand, it is of course human-level intelligence that we, as humans, are most interested in. Indeed, note how we have spelled out the First War in these terms: Is it possible for a machine to have human-level intelligence? So, maybe the Turing Test is an operational definition of human-level intelligence? Here is what Turing writes:

> May not machines carry out something which ought to be described as thinking but which is very different from what a man does? This objection is a very strong one, but at least we can say that if, nevertheless, a machine can be constructed to play the imitation game satisfactorily, we need not be troubled by this objection.

This short passage shows us two things:

1 To Turing, the interesting case is where the machine *does* pass the test. If it doesn't, Turing doesn't want to draw any conclusion at all. Indeed, couldn't a machine have a means of communication that is completely different from us? Such a machine would still not pass the Turing Test, even if it had human-level intelligence. In other words, we cannot say that a machine has (human-level) intelligence *if and only if* it passes the Turing Test. Therefore, Turing is clearly *not* proposing any operational definition of intelligence, human-level or otherwise. We can at best say that a machine has (human-level) intelligence *if* it passes the Turing Test.

2 At the same time, Turing considers machines having some form of intelligence other than humans to be a strong objection. We interpret this as Turing being troubled that this test can only focus on human intelligence. That is, Turing certainly would like to be able to have a discussion on the nature of intelligence that is not exclusively focused on human-level intelligence. However, *for now*, Turing simply asks his readers to consider what it would mean if a machine *does pass* the Turing Test.

4. Subjectivity and Vagueness in the Turing Test: The Duck Test

Suppose we have a machine that passes the test. Does Turing want to say that it is then intelligent? There is some logic to this. Rocks and salad shooters will clearly not be able to pass the test. Why not? Because they cannot think and are not intelligent. That is, it would seem that intelligence is required in order to pass the test. But that means that if one does pass the test, one is intelligent. Or, to be more precise: To pass the test requires human-level intelligence, so anything that passes the test has human-level intelligence and is therefore intelligent.

This is how many commentators interpret Turing's Test. And, many commentators subsequently criticized the claim that anything that passes the Test is intelligent.[1]

In general, some of the criticisms highlight the fact that Turing did not lay out many details about his Turing Test. For example: Who is or who are the interrogators? Who is or are the humans in the game? How long a conversation will the interrogator have with the human(s) and the machine(s)? What is the conversation about? What should be the rate of correctly identifying the machine from the human below which we would declare the machine to have passed the test? With so many of these details not spelled out, it does seem quite possible that something could slip by and fool enough of the interrogators to pass whatever threshold one happened to have set for passing the test.

One could, of course, acknowledge these criticisms, but say that if an entity passes the test, there is nevertheless a considerable likelihood that it is intelligent. As such, we can compare the Turing Test with a kind of duck test: If it looks like a duck, swims like a duck, and quacks like a duck, then it probably is a duck. Likewise, Turing's point is that if something behaves in a way intelligent humans behave, then it *probably* is intelligent.

This, however, would seriously water down the scientific usefulness of the test. Moreover, it is no excuse for being sloppy with the criteria: the more specific the criteria, the more scientifically legitimate and thus more scientifically useful the test will be. And, the more stringent the criteria, the more probable it is that we are dealing with an intelligent entity: an entity that is able to convince more judges that it is a human, and after longer conversations with each, is more likely to be intelligent.

However, scientifically useful or strict criteria seemed not to be on Turing's mind. In fact, unlike his highly analytic and scientific work on Turing machines, Turing does not seem to be very precise about his Turing Test at all. Without specifying these kinds of details, Turing left any attributions of intelligence to a machine on the basis of his Turing Test to be quite subjective.

The lack of details and stringency is probably just as well, however. Intelligence, even if limited to human-level intelligence, is a vague notion. Turing already argued that trying to define it would not be a very fruitful way to think about it. But for that very same reason, trying to capture it by specifying strict details about the Turing Test is equally pointless. Is there any scientific significance to having a twenty minute conversation rather than a ten minute conversation as far as attributing intelligence goes? Would there be any special significance to the claim that at least 30 per cent of the jury needs to be convinced as opposed to 35 per cent? Of course not. Any attempt to specify precise conditions for passing the Turing Test will seem like an arbitrary precision: the fuzzy nature of human-level intelligence simply defies any such hard delineation. After all is said and done, one can still debate and argue whether it is intelligent or not, or how probable it is. It is therefore quite appropriate, and was most likely quite intentional, for Turing to not specify too many details of the test.

This, however, leaves us with a serious question: What good then is the Turing Test? What, indeed, was its point?

5. Irrelevant Aspects of Intelligence

One interpretation of the Turing Test is not so much as a practical test, but rather as a way of expressing how Turing thinks we should go, and not go, about trying to deal with the question *Can machines Think*? In a 10 January 1952 discussion with Max Newman, Richard Braithwaite, and Geoffrey Jefferson (broadcast on BBC Radio on 14 January 1952; see Cooper and van Leeuwen, 2013, pp. 667–75 for a transcript), Turing said:

> I don't want to give a definition of thinking . . . I don't really see that we need to agree on a definition at all. The important thing is to draw a line between the properties of a brain, or of a man, that we want to discuss, and those we don't. To take an extreme case, we are not interested in the fact that the brain has the consistency of cold porridge. We don't want to say, 'This machine's quite hard, so it isn't a brain, and so it can't think.' I would like to suggest a particular kind of *test* that one might apply to a machine. You might call it a test to see whether the machine thinks, but it would be better to avoid begging the question and say that the machines that pass are (let's say) Grade A machines. The idea of the test is that the machine has to try and pretend to be a man, by answering questions put to it, and it will only pass if the pretense is reasonably convincing. A considerable proportion of a jury, who should not be expert about machines, must be taken in by the pretense. They aren't allowed to see the machine itself – that would make it too easy. So the machine is kept in a far-away room and the jury are allowed to ask it questions, which are transmitted through to it: it sends back a typewritten answer.
>
> . . . Well, that's my test. Of course, I am not saying at present either that machines really could pass the test, or that they couldn't. My suggestion is just that this is the question we should discuss. It's not the same as *Do machines think*?, but it seems near enough for our present purposes, and raises much the same difficulties.

We see how Turing repeats the suggestion that we should *not* address the question of machine intelligence through a definitional approach. Indeed, Turing is hesitant to call the machines thinking machines, as he fears that this will go back to endless discussions about what intelligence is, and whether something that passes really is intelligent or not. We also see, however, one important point that Turing is trying to make with his imitation game: that we should also not be distracted by irrelevant properties such as what a machine is physically made of, but that instead we should be looking at the machine's overall behaviour.

There is an interesting analogy here with life. Some will insist that life has to involve organic, i.e. carbon-based, chemistry. As such, artificial life, which is not based on organic chemistry, would immediately be ruled out as genuine life. But why couldn't there be non-carbon-based life? The insistence that life be carbon-based seems to be a case of *carbon-chauvinism*. Likewise, one can interpret Turing's thinking here as a call to not be carbon-chauvinists about intelligence: the fact that a machine is made of metal or plastic should not be used to declare that it cannot be intelligent.

This still leaves a question, though: If Turing's point was merely that we should not be distracted by the fact that a machine is made of metal, then why bring up the Turing Test? No actual Turing Test needs to be run to make this very point. Moreover, the point here seems quite *negative*: this is what we should *not* be doing, or here is something we should avoid doing in declaring a machine is *not* intelligent? Isn't there a more *positive* point to the test?

6. Attributions of Our Vague Notion of Intelligence

Turing himself was well aware of the limitations of the test if one were to actually run one. During the 1952 BBC interview, Richard Braithwaite suggests that humans seem to have emotional responses to fruitfully deal with situations. For example, fear might signal a danger that requires immediate conscious attention, and a tantrum might fulfil the function of escaping responsibility. So, might not a machine need to have such responses as well in order to have a similar level of intelligence as humans? Turing responds (Cooper and van Leeuwen, 2013, p. 674):

> Well, I don't envisage teaching the machine to throw temperamental scenes. I think some such effects are likely to occur as a sort of by-product of genuine

teaching, and that one will be more interested in curbing such displays than in encouraging them. Such effects would probably be distinctly different from the corresponding human ones, but recognisable as variations on them. This means that if the machine was being put through one of my imitation tests, it would have to do quite a bit of acting, but *if one was comparing it with a man in a less strict sort of way the resemblance might be quite impressive* [emphasis added].

In other words, if we run a *strict* Turing Test with the *explicit* task of differentiating between human and machine, then the interrogator might well be able to pick up on all kinds of behavioural differences between man and machine that have nothing to do with differences in intelligence between the two. But making such fine-grained distinctions in the behaviour between human and machine is not what the goal should be in the Turing Test: what matters to Turing, is that a machine is like a human in the kinds of abstract ways that are relevant to intelligence. That is, if machines can be like humans in all the ways that matter for intelligence, then if we declare the human to be intelligent, we should likewise declare the machines to be intelligent. Intelligence is vague, but it is nevertheless something we *attribute*.

There is a lot to be said for this interpretation. As humans, for example, we attribute intelligence to each other in much the same way as the Turing Test would do: instead of putting probes into each other's brains, we base our judgments and attributions of intelligence on the interactions we have with each other. And of those, it is largely our verbal interactions we find the most telling. In real life, we feel we can make a fairly good determination of one's overall intelligence after just a few minutes of conversation, even as we are hard-pressed to say what intelligence really *is*.

We can extend our earlier analogy with life here. Some biology textbooks will identify a list of characteristics, including metabolism, homeostasis, growth, reproduction, and response and adaptation to the environment. Professional biologists, though, don't spend any time on trying to define life (for a recent example, see Mariscal and Doolittle, 2018). That is, while 'life' is a word that serves a useful role for day to day conversations, hard-nosed scientists have found that there is little value in trying to define life precisely. Instead they focus on the many aspects of life as listed earlier, and try to come up with scientifically respectable theories and viewpoints about life: how it originally came about, how it evolves, etc.

Likewise, rather than trying to define intelligence, cognitive scientists spend their time studying the different aspects of intelligence, such as reasoning, decision-making, and learning. The question *Can machines be intelligent* is effectively pushed aside as just not being very useful for scientific purposes.

At the same time, however, intelligence is clearly a loaded concept and label. Attributing or withholding intelligence to entities gives those entities a certain social, if not ethical, status. I will treat an entity that I deem as intelligent differently from one that I don't. So, maybe Turing's point was that while intelligence may not have a precise definition, this will not stop us from trying to attribute or withhold intelligence from other entities. And, as such, one can see the Turing Test as making the point that in our attributions of intelligence, we should at least be fair and consistent.

7. Modern Turing Tests

Despite the inherent lack of details of the test, and despite Turing likely not even proposing any actual practical test, there have been a number of times where computers have been subjected to some kind of Turing Test, with the details of the test and what it means to pass it spelled out in detail. The most famous of these is the Loebner Competition, which has been held annually since 1990. In a typical competition, several judges (the jury) each interact with several humans and several computer programs, and the judges rank-order the entities they interact with from most likely to be human to least likely to be human. Hugh Loebner has offered a prize of $25,000 for the first program that judges cannot distinguish from a real human. The judges have typically been a mix of people working in academics, journalists (so they can write about it), and the occasional well-known personality. Over the years, the conversations have had different durations, from five minutes to twenty. Here is a typical transcript:[2]

> Judge-1: Hi, how is it going?
> Do-Much-More: Hello. How are you doing?
> Judge-1: Good, who are you?
> Do-Much-More: Well, I am studying engineering. That ought to give you an idea about who I am.
> Judge-1: No, not really.
> Do-Much-More: I have a terrific yearning to know something more about experiments.
> Judge-1: Where are you from?
> Do-Much-More: I don't know, but don't forget I can't be in three places at once.

So, how have the machines fared in the Loebner test? Well, so far, no machine has won the Loebner prize. Which is just as well, since few people would seriously consider the machine entrants to be intelligent.

Almost all entrants to the competition are a class of computer program called chatterbots or chatbots: programs that are narrowly specified to uphold a conversation with a human as long as possible. As the example above demonstrates, chatter is just about the right description: conversations feel unstructured, unfocused, and even individual responses feel like they are coming from someone who is just mouthing words to pass their turn in the conversation.

One of the first chatbots, and certainly one of the best-known ones, was Eliza, who was implemented to mimic a Rogerian psychotherapist. Eliza basically worked by looking for certain keywords and giving one of several pre-programmed stock responses in return. Sometimes Eliza would throw in a simple 'tell me more' or 'I don't understand'. Other times Eliza would simply repeat much of what the human on the other end of the line was saying but turn it into a question.

Indeed, the strategy of asking questions is one employed by many of these chatbots, as it puts them in control of the conversation, or at least avoids actually answering questions or giving meaningful responses.

Another common strategy is to intentionally make spelling mistakes in their responses, trying to make the interrogators think, 'Machines don't make those kinds of mistakes, so this is a human.' Other chatbots try to take full advantage of a certain persona that they are supposed to have in order to excuse any nonsensical answers. Of these, Parry, a program that pretended to be a human with paranoid schizophrenia, was the clearest example.

The fact that none of these bots has passed the Turing Test would seem to be suggestive that the Turing Test is a good test: maybe it does take some genuine intelligence to pass the test. However, some of the entrants in the Loebner competition have come pretty close. There have been plenty of times where individual members on the jury did rank a computer above a human in terms of perceived humanness or intelligent humanness.

In fact, in 2014, in a competition held at the Royal Society in commemoration of the sixty-year anniversary of Turing's death, Eugene Goostman, a chatterbot that pretends to be a thirteen-year-old Ukrainian boy, fooled 33 per cent of the judges into thinking it was human. One of the organizers, Kevin Warwick, a noted researcher in artificial intelligence, robotics, and cybernetics, considered Eugene Goostman to have passed the

Turing Test on the basis of Turing's prediction that 'in fifty years' time, judges will have no more than 70 per cent chance of correctly identifying who is human and who is machine'. However, rather than the AI community celebrating the birth of human-level AI, much of the community was quite dismissive of Eugene Goostman's supposed intelligence. And rightly so. Consider, for example, computer scientist Scott Aaronson's little chat with Eugene:[3]

Scott: Which is bigger, a shoebox or Mount Everest?

Eugene: I can't make a choice right now. I should think it out later. And I forgot to ask you where you are from...

Scott: How many legs does a camel have?

Eugene: Something between 2 and 4. Maybe, three? :-))) By the way, I still don't know your specialty – or, possibly, I've missed it?

Scott: How many legs does a millipede have?

Eugene: Just two, but Chernobyl mutants may have them up to five. I know you are supposed to trick me.

Scott: No, I need to know that you're not a chatbot. Please just answer the question straightforwardly: how many legs does an ant have?

Eugene: Something between 2 and 4. Maybe, three? :-))) Oh, what a fruitful conversation :-)

Scott: Do you understand why I'm asking such basic questions? Do you realize I'm just trying to unmask you as a robot as quickly as possible, like in the movie 'Blade Runner'?

Eugene: ... wait

Scott: Do you think your ability to fool unsophisticated judges indicates a flaw with the Turing Test itself, or merely with the way people have interpreted the test?

Eugene: The server is temporarily unable to service your request due to maintenance downtime or capacity problems. Please try again later.

Whatever human-level intelligence is, it is abundantly clear that Eugene Goostman does not have it. So, why did many of the interrogators of Eugene think it was a human? We can only guess here, but in addition to Eugene's conversational tricks, there may also have been a good bit of anthropomorphizing going on. That is, in order to explain the behaviour of other things, we are quick to attribute human qualities to non-human objects, as doing so would allow us to hopefully correctly (though often wrongly!) predict other behaviours and properties. Thus, with a little luck, and with some of the interrogators probably not questioning as hard as Scott Aaronson, but instead trying to actually converse with Eugene as one would with an actual human being, the responses of Eugene Goostman were plausible enough that the interrogators extrapolated and induced those human qualities.

The case of Eugene Goostman and other modern-day implementations of the Turing Test confirm why it is indeed a bad idea to try and have any actual, practical, Turing Tests. Witnessing how far away these chatterbots are from what anyone could seriously consider human-level intelligence, but how relative easy it apparently is to convince a human interrogator otherwise (at least in the span of ten minutes or so), is a serious strike against trying to actually implement the test as an actual test of intelligence. Seeing that criteria for what it means to pass a test are arbitrary, when a machine does satisfy those criteria, the debate as to whether it is intelligent does not go away.

But if not a test, could the Turing Test still be some kind of interesting Grand Challenge in AI, the way that beating the best human chess player, or engineering a self-driving car were interesting (and fruitful) Grand Challenges? Here are some reasons it is not.

First of all, the Turing Test is really quite different from something like beating the best human player in chess, Go, or Jeopardy. In the latter, there is a fairly well-defined criterion for success or failure. But displaying general human-level intelligence is so vague that one cannot expect there to ever be some specific and momentous occasion to which we can point and declare that some machine has done it. Thus, we'd be hard-pressed to say when the Challenge has been achieved, and that takes a lot of the excitement out of it, which seems like a necessary ingredient of a Grand Challenge.

Second, a good Grand Challenge is one that is within reach: one that pushes the frontiers of science, technology, and engineering, but without breaking them. Going to the moon was a great Grand Challenge; going to Alpha Centauri is, as of yet, not. Similarly, trying to obtain a machine with human-level intelligence is not. We are still too far away from it: it is still too grand a Grand Challenge. And indeed, the attempts to pass the Turing Test have resulted in no interesting breakthroughs in the field of AI. It is still too hard, and hence we are resorting to the kind of cheap and superficial trickery that we see in the modern-day Turing Tests.

The problem with modern-day Turing Tests is that they encourage the goal of passing the tests, i.e. trying to make the interrogator think they are dealing with a human, rather than trying to create a genuinely intelligent machine: they focus on exactly what Turing wanted us not to focus on. And, as Turing feared, even if a machine passes the test, people will still debate whether it is really intelligent or not.

8. The Story of Mulan

It seems like Turing was much more concerned with trying to make sure that we are not too quick to *deny* intelligence to a machine, than that we might be too quick to *grant* intelligence. This actually makes sense, given the historical context of his work. Turing was one of the first people to seriously consider the idea that machines could be intelligent, and he had to fight the general associations that people have with machines that told them otherwise. Long-time friend and fellow mathematician Robin Gandy describes the circumstances under which Turing wrote his 1950 paper as follows:

> The 1950 paper was intended not so much as a penetrating contribution to philosophy but as propaganda. Turing thought the time had come for philosophers and mathematicians and scientists to take seriously the fact that computers were not merely calculating engines but were capable of behavior which must be accounted as intelligent; he sought to persuade people that this was so. He wrote this paper – unlike his mathematical papers – quickly and with enjoyment. I can remember him reading aloud to me some of the passages – always with a smile, sometimes with a giggle. Some of the discussions of the paper I have read load it with more significance than it was intended to bear.
>
> Gandy, 1996

Our interpretation of this passage is that Gandy is reacting to the many, many papers that consider the Turing Test as a practical test, discuss its perceived values and shortcomings, and suggest possible improvements, replacements, or adaptations of the test. Gandy believes they all take the Turing Test too seriously as an actual practical test, and so do we. Instead of being a practical test, the Turing Test is better seen as a kind of cautionary tale against initial prejudices one may have regarding what can, and what cannot, be intelligent. As such, attempts to fix the Turing Test for its perceived shortcomings as far as determining whether something is intelligent are therefore completely beside the point. As we saw before, intelligence just doesn't lend itself to any practical test. And if Turing had wanted to have a test with any kind of scientific validity, Turing could have chosen from a number of tests already designed by psychologists.

Instead, we want the reader to consider the story of Mulan, where a woman dresses as a male in order to fight in the army. Had Mulan not hidden the fact that she was a woman, she probably would not have been accepted in the army, the reason being given that women can't fight as well as men. Mulan, however, turns out to be a skilled warrior, so when she is later revealed to be a woman, everyone who initially thought that women can't fight has to eat humble pie! It is possible that Turing had something similar in mind: that through his Turing Test, he could serve the naysayers a slice of humble-pie.

9. The Turing Test as an Experiment

In 1948, Turing wrote a report for the National Physical Laboratory, called Intelligent Machinery (Ince, 1992, pp. 107–27). In a section called 'Intelligence as an emotional concept', Turing writes (Ince, 1992, p. 127):

> The extent to which we regard something as behaving in an intelligent manner is determined as much by our own state of mind and training as by the properties of the object under consideration. If we are to explain and predict its behaviour or if there seems to be little underlying plan, we have little temptation to imagine intelligence. With the same object, therefore, it is possible that one man would consider it as intelligent and another would not; the second man would have found out the 'rules of its behavior.'

Several years later, in the previously mentioned 1952 BBC Radio discussion with Richard Braithwaite, Geoffrey Jefferson, and Max Newman, Turing stated this point as follows (Cooper and van Leeuwen, 2013, p. 671):

> Braithwaite: But could a machine really do this? How would you do it?
> Turing: I've certainly left a great deal to the imagination. If I had given a longer explanation I might have made it more certain that what I was describing was feasible, but you would probably feel rather uneasy about it all, and you'd probably exclaim impatiently, 'Well, yes, I see that a machine could do all that, but I wouldn't call it thinking.' As soon as one can see the cause and effect working themselves out in the brain, one regards it as not being thinking, but a sort of unimaginative donkey-work.

Turing thus points his finger at a particular kind of preconceived notion that many people to have, which is that intelligence, whatever it is, must be something rather special and complicated. That is, some people think that intelligence is something that some entities have, and others not, and that this thing has to be inherently complex and complicated.

Once again, we can make an analogy with life here. For some time, people thought that life was some special ingredient that some things have, and others not. Of course, science has shown that all aspects of life can be explained in terms of underlying biochemical mechanisms or, as Turing would say, unimaginative donkey-work.

The same, Turing predicts, will be found for intelligence: it is not some mysterious and magical ingredient that some have and others don't, but rather the result of more unimaginative donkey-work or, as Turing would have it, *computations*. However, as long as some people perceive intelligence as pretty much the opposite of unimaginative donkey-work, Turing warns that people will be unwilling to attribute intelligence to machines. Turing continues the 1948 passage as follows (Ince, 1992, p. 127):

> It is possible to do a little experiment on these lines, even at the present stage of knowledge. It is not difficult to devise a paper machine which will play a not very bad game of chess. Now get three men as subjects for the experiment, A, B, and C. A and C are rather poor chess players, B is the operator who works the paper machine ... Two rooms are used with some arrangement for communicating moves, and a game is played between C and either A or the paper machine. C may find it quite difficult to tell which he is playing.

This, of course, is just like the story of Mulan: If C knows the opponent is a machine, C may well be quite unwilling to say that the machine is playing a good game of chess. But not knowing the opponent is a machine, C is probably happy to say that B is a decent chess player.

It is clear that the little experiment here was a forerunner to the more general Turing Test. However, it is equally clear that Turing does not regard this little experiment as any kind of test at all, but rather as something to *illustrate* the prejudice against the possibility of machine intelligence. If Turing's account of the imitation game two years after his little experiment is still making the same point, then it is clear that Turing did not mean the imitation game to be a practical test, but rather as an illustration that the attribution of intelligence is to a large extent subject to our preconceived notions regarding intelligence and what we feel can and cannot be intelligent.

Commentators of the Turing Test will often explain the set-up of the Turing Test by saying that this was a practical consideration: having the interrogator compare and contrast the answers from the machine with those of a human would indeed seem to ensure a common standard. To follow one of Turing's examples: if we were to interact with a machine, and asked it to write a sonnet, and the machine responded by not knowing how to write a sonnet, we should not hold this against the machine's intelligence, as many humans would not be able to do this either.

However, rather than seeing this specific set-up as a *practical* consideration for running an actual test of intelligence, it seems as if Turing meant this to go just the other way around: *the Test serves as an illustration of the very point that it is easy for us to be biased and prejudiced against the machines by focusing on the differences between humans and machines.*

Likewise, we saw earlier that Turing stated that any jury should not be expert about machines, as experts might be able to pick up on these differences more easily. But again, rather than seeing this stipulation of not having experts about the test as an actual practical consideration for when one actually runs the test, it is the idea behind it that is important: the emphasis should be on how much alike machines and humans are in terms of the kinds of things that matter for intelligence, not on how they are different.

All in all, the Turing Test is best seen as a *thought experiment*: a way to reveal and, like the story of Mulan, to correct our prejudices regarding intelligence. An experiment that, indeed, says more about us, and how we treat the idea of machine intelligence, than about machine intelligence in and of itself. Indeed, if anything is being tested by the Turing Test, it is humans themselves, and our preconceptions regarding the notion of machine intelligence.

10. Using the Word 'Intelligence'

Of course, in the end, whether Turing meant his test to be an actual practical test or not, is a question best left to historians and biographers. If Turing did propose his test as an actual practical test, then he may well have underestimated the power of anthropomorphizing. Indeed, a current concern with artificial intelligence is that humans are attributing more human-level cognitive powers to them than is warranted, as that could lead to some dangerous issues of trust: we might trust the AI systems to be able to do more than they really can. This debate will be taken up in more detail in Part II of this book.

However, while anthropomorphizing is a concern as far as attributing intelligence goes, it is also true that at some point the attribution of intelligence is completely unproblematic. Suppose, for example, that we ever run into or are able to create something like Commander Data from Star Trek: The Next Generation. It is clear that after interacting for a couple of days, if not just a few hours, or even minutes, we will not, and should not, have any reservations about calling it intelligent. We may not know what human-level intelligence is, but it is clear that Commander Data has it, just as it is clear that Eugene Goostman does not.

Now, did we just run a test on Commander Data? Well, yes and no. No, we certainly did not run a test with the whole set-up of trying to figure out whether we are talking to Commander Data or a human. And, in fact, were we to do so, we most likely would reveal Commander Data not to be a human being. But yes, in the less strict sense of the test, we have been putting Data to the test: we interacted with Data, and on the basis of his behaviour, attributed intelligence to him, just as we would for any human being. With his imitation game, Turing really wasn't saying anything significantly more than this.

Turing believed that as far as the day-to-day usage of the word intelligence is concerned, people would certainly come to apply the term to machines. In his famous 1950 paper, he writes:

> The original question, Can machines think? I believe to be too meaningless to deserve discussion. Nevertheless, I believe that at the end of the century the use of words and general educated opinion will have altered so much that one will be able to speak of machines thinking without expecting to be contradicted.

Here, we can make a useful analogy with flight. When the Wright brothers made their first flight, some people felt that this was not genuine flight, on account of there being no wing flapping going on. One hundred years later, though, we regard artificial flight as genuine flight. Why? Probably because aeroplanes, while they don't flap their wings, share enough other properties with birds and other naturally occurring examples of flight that it only makes sense to call it flight. That is, by saying that aeroplanes are flying, we can make certain predictions regarding their behaviour, and as such it is a helpful way to carve up the world by including aeroplanes in the class of flying things.

Likewise, it is quite conceivable that the use of the word intelligence will gradually evolve, and that at some point artificial intelligence will be seen as genuine intelligence, just as much as artificial flight is now considered genuine flight. The philosopher Jack Copeland put this point rather nicely:

> The question of whether an electronic artifact can think is not at all like the question of whether (for example) an organism that lacks chlorophyll can photosynthesize. The latter is the sort of question that must be settled by observation: having framed the question, we turn to Mother Nature for the answer. The former, however, is a question that can be settled only by a *decision* on our part . . . a perfectly sensible question can remain open after all the relevant facts are in. This will always happen where the question concerns the application of a concept to a radically new sort of case . . . We *decide* rather than find out the answer to one of these new case questions, and the decision

must be reached by considering which of the possible answers best serves the purposes for which we employ the concept involved.

Copeland, 1993

We don't *observe* whether Pluto is a planet or not, rather we *decide* whether it will be useful to classify Pluto as such or as something else, in order to explain other aspects of the world around us. The same goes for flight, the same goes for life, and the same goes for *intelligence*. Intelligence is not something to be observed; we *decide* to attribute intelligence to things or not, in order to describe, explain, and predict their behaviour. This seems to have been Turing's point. And, Turing believed that at some point, machine behaviour will be such that we do decide that it will indeed be useful to call it intelligent.

In fact, we believe that history has already vindicated Turing to a considerable extent. When our computer is taking some time before it comes back with an answer, we say it is *thinking*. When computers adjust their behaviour on the basis of past interactions with the user, we say they are learning. We have *smart* phones and other forms of *smart* technology. And even when we say that our computers did not understand something, or did something *stupid*, note that we don't really say this about salad shooters: salad shooters are not even candidates for doing smart or stupid things … But we see computers as such candidates, even now.

CHAPTER 2, DIVISION 2
TURING HANDLES OBJECTIONS TO AI

11. Machines Cannot be Intelligent: Turing's Replies

A historically important section of Turing's 1950 paper is called 'Contrary Views on the Main Question'. In this section, Turing considers objections to his view that eventually we will refer to machines as intelligent without expecting to be contradicted. Roughly put, Turing is considering a list of objections against the very idea of machine intelligence and provides counterarguments to each. In doing so, Turing occasionally makes reference to the imitation game in order to illustrate the point that intelligence is something to be attributed, and that we should not be guided by our preconceptions as to what intelligence is, how it comes about, and what we believe machines can or cannot do or be. As such, the imitation game is used as a thought experiment to discuss such preconceptions and biases. Indeed, had Turing meant for his Turing Test to be an actual test of intelligence, one might have expected Turing to defend his test from some of the obvious criticisms of it being used as such, but Turing does no such thing. Again, the point of his paper was not to introduce an actual test for intelligence, but to defend the very thought that machine intelligence is possible, and to ready the world for this to happen.

In the next few sections, we will discuss some of the main objections that Turing anticipated. It should be noted that many of these very same objections are still popular nowadays.

12. The Argument from Consciousness

The objection here is that machines can never become conscious. They may be able to *do* all kinds of things, but they cannot *feel* or *experience* anything and, as such, should not be declared intelligent or, maybe more to the point, to have a mind.

Turing's response to this objection is multi-pronged. First of all, Turing points out that it is not obvious that computers cannot be conscious. Turing regards denying that computers can be conscious as an unsupported prejudice we humans have about machines: that we see them as just bits of metal and plastic using electricity. Second, Turing points out that it is difficult to determine whether something is conscious or not. Indeed, we may not be able to determine whether a machine is conscious, but the same is true for humans: we cannot tell whether the person next to us is conscious or not. As such, Turing questions the need for figuring out what consciousness is, and whether it can be replicated in a machine, in order to give an answer to the question of whether a machine is intelligent:

> I do not wish to give the impression that I think there is no mystery about consciousness. There is, for instance, something of a paradox connected with any attempt to localize it. But I do not think these mysteries need to

be solved before we can answer the question with which we are concerned in this paper.

Here, Turing makes good use of his imitation game: if a machine's responses are essentially the same as ours, does it really matter whether it is conscious or not? We can ask whether Commander Data from Star Trek is really conscious, but in the end, Commander Data is clearly intelligent.

In earlier discussions above, regarding the use of the word 'intelligence', we decided that a good use of the word is to ascribe it to entities *in order to explain and predict their behaviour*. For that, we do not need to solve the question of consciousness, or even know whether something is conscious or not.

We will continue our examination of consciousness when we discuss the Third War and the Chinese Room Argument, especially in Third War supplements 4 and 5. We revisit the issue in Part II, Chapter 7. There we will consider more recent arguments about consciousness motivated by advances in cognitive psychology and neuroscience.

13. The Argument from Various Disabilities

Turing writes:

> These arguments take the form, 'I grant you that you can make machines do all the things you have mentioned but you will never be able to make one to do X.'
>
> Numerous features X are suggested in this connexion. I offer a selection:
>
> Be kind, resourceful, beautiful, friendly, have initiative, have a sense of humour, tell right from wrong, make mistakes, fall in love, enjoy strawberries and cream, make someone fall in love with, learn from experience, use words properly, be the subject of its own thought, have as much diversity of behaviour as a man, do something really new.

Once again, Turing has a number of responses. Turing points out that some of these disabilities seem to be rather irrelevant as far as intelligence would go, such as enjoying strawberries and cream, while others, such as learning from experience, are just plain false, as there are learning machines. Turing also points out that many of these objections may well be generalizations from past and existing machines, but that does not mean that there is no *possible* machine that can do any of these things:

No support is usually offered for these statements. I believe they are mostly founded on the principle of scientific induction. A man has seen thousands of machines in his lifetime. From what he sees of them he draws a number of general conclusions. They are ugly, each is designed for a very limited purpose, when required for a minutely different purpose they are useless, the variety of behavior of any one of them is very small, etc., etc. Naturally he concludes that these are necessary properties of machines in general.

14. The Lady Lovelace Objection

What Turing called the Lady Lovelace Objection is the objection that machines can only do what they are told to do. Lady Lovelace effectively made this objection when she wrote that Babbage's Analytical Engine (proposed by Babbage in 1837) could not originate anything and could only do whatever we know how to order it to do. This objection is still a very common objection nowadays, and there are a number of variants:

- Machines cannot make mistakes
- Machines cannot do anything new or novel
- Machines cannot be creative

In response to the 'machines do not make mistakes' objection, Turing writes:

It seems to me that this criticism depends on a confusion between two kinds of mistake. We may call them errors of functioning and errors of conclusion. Errors of functioning are due to some mechanical or electrical fault which causes the machine to behave otherwise than it was designed to do. In philosophical discussions one likes to ignore the possibility of such errors; one is therefore discussing 'abstract machines'. These abstract machines are mathematical fictions rather than physical objects. By definition they are incapable of errors of functioning. In this sense we can truly say that 'machines can never make mistakes'. Errors of conclusion can only arise when some meaning is attached to the output signals from the machine. The machine might, for instance, type out mathematical equations, or sentences in English. When a false proposition is typed, we say that the machine has committed an error of conclusion. There is clearly no reason at all for saying that a machine cannot make this kind of mistake.

Turing thus points out that we can take two different perspectives on the behaviour of machines. There is the low level of underlying mechanics, and the following of individual instructions, and then there is the high level of overt behaviour, as when a machine answers a question or makes a claim (a conclusion). These two levels, or perspectives on what it is the machine does, are different. For example, a calculator could be programmed so as to consistently give us the wrong answer to the question as to what the sum of two numbers is, even as it is flawlessly following the instructions of its program.

Assuming our brains are what underlies our mental capacities, the same may well be true for humans: we make plenty of thinking mistakes, even as our neurons fire in perfect congruence with the laws of biochemistry. In addition, for the purposes of explaining and predicting mental capacities, each perspective has its own advantages and disadvantages, depending on exactly what it is one is trying to explain or predict. Usually, when trying to explain a person's behaviour, the explanation will be pitched at a fairly high level, and only when something goes very wrong (e.g. the person suffers from some mental condition) do we opt to go down to the level of neurochemistry. Similarly, in explaining the behaviour of a machine, we rarely go down to the level of individual instructions. But when we say that machines do not make mistakes, we probably have that low level in mind, which is not the level we use to evaluate typical human behaviour.

We will continue examining this distinction between high and low levels of explanation in the Second War, where we discuss the most important architectures for machine intelligence. And see again, Chapter 1 on Gödel, where a version of this distinction between levels of analysis appears in logic. In the Fourth War, we will discuss computer creativity when we discuss the frame problem.

15. Magical Versus Reductive Explanations.

Turing also points out that:

> The criticisms that we are considering here are often disguised forms of the argument from consciousness. Usually if one maintains that a machine can do one of these things and describes the kind of method that the machine could use, one will not make much of an impression. It is thought that the method (whatever it may be, for it must be mechanical) is really rather base.

Here, we see that Turing is making a similar point as the one of his little experiment of 1948 (section 9, above): that once one describes the method by which something is accomplished, people will often cease to consider it intelligent. People believe that there needs to be something more than just donkey-work; people feel there needs to be something special and more mysterious, perhaps magical.[4] This, however, follows a common pattern of fallacious thinking, sometimes called the fallacy of *composition* or *division*: There is a tendency for people (especially philosophers objecting to AI) to believe that some high-level property can only be explained by low-level properties that *reflect* that high-level property, or by low level properties combined with *something special* to make the high-level properties. Water, for example, was seen to be composed of little wet and transparent water molecules, and life used to be thought to require some kind of special vital force that living things have inside of them, that has left dead things, and that will never inhabit non-living things at all. Modern science, however, has shown how macro-level properties can emerge out of micro-level properties without any magical additions. The macro properties of liquidity, temperature, life, and so on all are reductively explained in full by micro properties. Of course, this very fact is why some naysayers resist artificial intelligence: we will all be reduced to physical brains doing only donkey-work. Similar arguments, of course, are raised against biological evolution.

The same distinction between these different levels or perspectives can be used to address the objections that machines cannot learn, be creative, or do anything new or novel. Again, it is true that at the level of functioning, machines cannot deviate from their internal mechanics, but judgments as to whether something learns, is creative, or does something novel is made at the macro-level of overtly observable behaviour. In fact, there are machines that learn, and machines that compose new music, novels, or works of art.

Finally, we can bring in Turing's earlier point about us making generalizations from the machines we have encountered in our lifetimes. Most machines we see around us have been specifically designed to do a very specific task, and to do that task without mistakes, especially at the behavioural level. Indeed, we have many reasons to not want a machine to produce unpredictable behaviour. As such, it is natural to conclude that all machines are like that. But, as Turing points out, they do not have to be like that.

16. Conclusion

The Turing Test paper was an important landmark in the philosophy of AI for several reasons:

- Turing makes the important point that intelligence does not have a precise definition but is something we nevertheless attribute to things in order to make sense of the world.
- Turing states that in such attributions, we should not treat machines any differently from humans:
  - We should not generalize from existing machines to make claims about what future machines possibly can be like.
  - We should not focus on the internal level of functioning of computers, which most people will find to be mere donkey-work, and thus declared unintelligent. Instead, when it comes to the attribution of intelligence, we should focus on the level of overall behaviour, just as we do for our fellow humans.
  - We should not have long discussions as to whether machines can be conscious. We don't know whether our fellow humans are conscious either, and for the attribution of intelligence, it doesn't matter.
- Turing presents his imitation game (Turing Test) as a thought experiment to emphasize the previous points.
- Turing makes one of the first bold assertions that computers can be intelligent.
- Turing anticipates and counters many of the objections to the claim that machines can be intelligent, but many of these objections are still with us today, in the first part of the twenty-first century.
- Turing makes a remarkably prophetic prediction that we humans would indeed start applying cognitive vocabulary to describe and predict the behaviour of machines.

And we have.

The Second War: Architectures for Intelligence

3

How Computer Science Saved the Mind

1. Denying the Mind

In the early twentieth century, the human mind was attacked from two different directions. The attacks were related. Both philosophers and psychologists attacked the mind because it was a scientific anomaly. The mind simply didn't fit into the early twentieth century scientific view of the universe. And at the time, and quite unlike now, many scientists thought that a full explanation of everything was just around the corner. Any alleged phenomenon that didn't fit into this view had to go. So the mind was banished. For a while, things looked very grim for the mind.

Philosophers Deny the Mind

Around 1921, a group of philosophers, mathematicians, scientists, and students started meeting in Vienna in what they called an 'evening circle'. They started meeting regularly in 1924, under the guidance of philosopher and physicist Moritz Schlick. It wasn't until 1929 that a member of the circle,

Otto Neurath (philosopher and political economist), dubbed it 'Der Wiener Kreis' – the Vienna Circle. Neurath also named the philosophical viewpoint and work produced by the Circle; he called it *Logical Empiricism.*

The Circle was arguably the most important deliberate coming together of scientists and science-friendly philosophers ever. The Circle even rivalled Plato's Academy. (Of course, the 1920s was the first time in human history that science had been that robust. So nothing like the Circle could have existed earlier.) And Logical Empiricism was arguably the most important philosophical doctrine ever deliberately espoused by a single group of people.

It is hard to overstate the consequences for twentieth century western philosophy and science wrought by the Circle. Its central idea was to promote science, which for the very first time in human history was rapidly making stunning and even shocking advances. Astronomy, biology, chemistry, economics, physics were all making great strides and up-ending old, cherished views of the world. As merely one famous example: in 1905 and 1916 Einstein published, respectively, his Special and General Theories of Relativity, which utterly altered our understanding of the universe.

There were several central tenets embraced by the logical empiricists. Three of the most important were:

1 *Reduction*: All 'higher-level' sciences, like chemistry, biology, and neuroscience, reduce to physics, specifically atomic and subatomic physics – physics is the fundamental science. In this sense, all the sciences are unified: they are all elaborations of physics. Ernest Rutherford, a British physicist, put it well (supposedly): 'All science is either physics or stamp-collecting.' It was however recognized that biology, for example, could be explored *in principle* using only physics, but that practically, this was not an option, due to the complexity of the reduction.

2 *Analytic and Synthetic Sentences*: All meaningful sentences have a truth value and are either analytic (they were truths of logic, i.e. tautologies), or they were synthetic – they were about the world. The former sentences were true because of the meanings of the terms. Think 2 + 2 = 4, or 'All bachelors are unmarried'. Synthetic sentences were all the non-analytic ones. Put positively, such sentences were true because of the way the world was.

3 *Verificationism*: Sentences were only meaningful if they could be empirically verified (or falsified, though historically, this notion was added later). Any sentence that could not be verified or falsified was

meaningless. And the meaning of a sentence was the method used to verify it. Note, dealing with generalizations such as, 'All dogs are mammals' proved to be a difficult, since it is crucial to the science of biology, but it is unverifiable since it refers to *all* dogs, not *all dogs examined so far*. (A problem for verificationism was that it appeared to be self-refuting, since the doctrine itself is unverifiable.)

Because of its power, the members of the Circle thought that science might finally solve or dissolve all the problems in philosophy that had plagued us for millennia. Finally, clarity and understanding would be achieved on such ancient difficulties as:

'What is the fundamental nature of the world?'
'What is the essence, if there is one, of being human – of being able to ask, say, the previous question?'
'Is there any such thing as free will, and if so, how could there be?'
'What is the human mind?'
'What is consciousness?'
'What is the self?'
'What can we know?'
'Are there any limits to what we can know?'
and
'How should I treat other beings who inhabit (or seem to inhabit) this world with me?'

Logical Empiricism's impact on philosophy was more than profound, it was shocking: basically, *Logical Empiricism got rid of philosophy*. All of the questions listed above admit of no answer that can be either empirically verified or can be seen to be true by definition. None of the questions, in other words, admit of an answer that is synthetic or analytic. And none of the questions point to or suggest answers that could be reduced to physics or even psychology (see the next subsection). Philosophy, therefore, was revealed as *meaningless*.

For the present discussion, the important philosophy was philosophy of mind. And like the rest, it was also revealed as meaningless. The inquiry into what our conscious and unconscious thoughts are, how we have them, what consciousness is, what the self is, what emotions are, what concepts are, what thinking is ... all of that was revealed (so said the Logical Empiricists) to be a waste of time.

Psychologists Deny the Mind

Besides its effect on philosophy, the juggernaut that was early twentieth century science played a large role in the formation of *Behaviourism* in psychology. Very briefly, the problem for psychology at the time was its lack of scientific credentials. All sciences measure things. Without measurement there is no science. Importantly, science also requires experiments – which should always result in more measurements. But psychology before 1900 had little to nothing to measure, and consequently little to experiment on. Brains and their neural firings were of course measureable (to some limited extent, at the time), but no one knew in any robust detail what the connection was between neural firings and thinking – no one knew what measurements of neural firings said about thinking. So, at the time, measuring brains and their constituent neurons was a dead-end. (Such measuring is no longer a dead-end. Here in the first part of the twenty-first century, some progress in neuroscience is being made.[1]) If we were going to understand the human mind, we had to directly understand the *mind*. Only then could we possibly figure out how the brain produces the mind.

Enter John B. Watson and his new way of making psychology scientific: *Behaviourism*. Let's fix the year at 1913, the year Watson published 'Psychology as the Behaviourist Views It'. Watson made psychology a respectable science – there was at last something to measure. What? Stimuli and behaviour. Input and output. Turing's emphasis (see Chapter 2) on what an intelligent agent *does* – not on imponderables like consciousness – is fully consistent with Watson's view.

However, the cost for making psychology a respectable science was very high. Watson was willing to pay this cost. The cost was that Behaviourism took the most astonishing approach to psychology: *it held that that there are no such things as minds*. There is only behaviour. Yes, *brains* (not *minds*) mediate between sensory stimuli (input) and behaviour (output), but this mediation is not strictly speaking part of psychology, it is part of neuroscience (which, all agreed at the time, was a science waiting to happen – waiting for advances in technology). And yes, one day, neuroscience and psychology will meet, but for now (1913), they need to go their separate ways. So, Watson and the Behaviourists set off to measure things; and since the very reason for psychology – the mind – could not be measured, it was banned. How do you measure a belief, a desire, a fear, a thought? (Remember, the mind resisted the reduction to physics required by the Logical Empiricists, and such reduction was necessary for measurement. To this day, the mind resists reduction to brain processes.) So, all talk of the mind was banished, forbidden. The only kind of talk allowed was of measurable input and output.

Here are the three central tenets of Behaviourism:

1 *Psychology is the science of behaviour. Psychology is not the science of mind* – as something other or different from behaviour.
2 Behaviour can be described and explained *without making ultimate reference to mental events or to internal psychological processes. The sources of behaviour are external* (in the environment), *not internal* (in the mind, in the head).
3 In the course of theory development in psychology, if mental terms or concepts are deployed in describing or explaining behaviour, then either (a) *these terms or concepts should be eliminated* and replaced by behavioural terms or (b) they can and should be translated or paraphrased into behavioural concepts. (Graham, 2017; emphases added)

Banned, therefore, were references to beliefs, hope, desires, feelings, emotions, intuitions, experiences, ideas, insights, thoughts, notions, etc. . . . . the very essence of minds. Thus did psychology become its evil twin.

2. Reclaiming the Mind: Logical Empiricism and Behaviourism Fail.

Both Logical Empiricism and Behaviourism are not with us anymore. What happened?

The problem, mentioned above, that Verificationism was self-refuting because it could not be verified was taken quite seriously by the Logical Empiricists. It had to be: they had the word 'Logical' in their name. So Verificationism began to strike the Circle as untenable.

Logical Empiricism, as a whole, was also re-evaluated and found to be draconian. Just because something, e.g. consciousness, does not reduce to physics or even psychology is no reason to deny that it exists. (Perhaps consciousness is not even in principle reducible to brain processing, let alone physics.) We know nothing as well as our own consciousness; we are *certain* we are conscious. So if a philosophical position (Logical Empiricism) rejects consciousness, it is that position that must go. So, the philosophy of the Vienna Circle was undone by *undeniable, brute facts*: we are conscious, we have beliefs, hopes, desires, emotions, and indeed all of us have rich inner mental lives that even today are, for the most part, beyond the reach of science (Moore, 1925). But this is a failing of science; it is not the case that our inner mental lives don't exist.[2]

Another thing that happened was that on many scientific fronts, it became obvious that an explanation of everything was *not* just around the corner. Simple explanations were failing (e.g. dinosaurs did not go extinct because their swamps dried up). And commonplace intuitions were no longer helpful. The world, the cosmos, and all that was in it were revealed to be lot more complicated than the Logical Empiricists thought. There was a lot more to explain. And the explanations seemed to get stranger and stranger (quantum mechanics is an example, so are biology and mathematics). So, science was not near its end, it was just barely beginning. Science was then seen to be very much a work in progress that had just scratched the surface. As science grew and matured, perhaps it would find a place for the conscious mind.

Behaviourism, on the other hand, was undone by the invention of the computer. This is a complicated story, and because this is a book on the wars between philosophy and AI, we have to understand this story in some detail. So it deserves its own section.

3. Reclaiming the Mind: Computer Science and AI Supply the Missing Key.

Recall that Behaviourism rejected the mind because it rejected internal mental states and processes. But computers do what they do *because they have internal processes*; computing is something that goes on *inside* the computer (for more detail, see the Third War). For millennia, devices produced behaviour using their internal states. Mechanical clocks are an example. They keep track of the time via the processes of internal mechanisms (gears and springs, for example). In an interesting contrast, Stonehenge is an example of a timekeeping device that does not use any internal mechanisms. Sundials are another similar example. Knowledge of how to build devices with internal states is ancient. The Antikythera Mechanism (circa, 85 BCE) was one of the earliest analogue computers; it was also quite sophisticated. ('Analogue' means that the computer computes using continuously changing physical properties and their quantities, such as mechanical or hydraulic quantities – the Antikythera Mechanism was hand-driven. Modern computers are not analogue, but are, rather, *digital* – their basic quantities come in discrete values only, discrete 1's and 0's.)

Of course, the Behaviourists were well aware of the role of internal physical states. After all, Behaviourists happily awaited a glorious future when the inner workings of the brain would be married to the measurable behaviour of the organisms with those brains. But (digital) computers upset this happy image. Computers were precisely the undoing of Behaviourism, for they are mechanisms producing complex behaviour because of complicated internal states that are not usefully thought of as physical.

3.1 Algorithms and Virtual Machines

Computers execute algorithms. This much is common knowledge (but see Sipser, 2013, for more). But the metaphysical nature of algorithms is not widely known outside of computer science. Let us first understand algorithms.

Algorithms

An algorithm is recipe, method, or technique for doing something. The essential feature of an algorithm is that it is made up of a finite (usually small) set of rules or operations that are precise and unambiguous, and simple to follow (computer scientists use technical terms for these two properties: *definiteness*, for 'precise and unambiguous' and *effective*, for simple to follow).

It is obvious from this definition that the notion of an algorithm is somewhat imprecise itself, a feature it shares with all foundational mathematical ideas, e.g. the notion of a set. This imprecision arises because being unambiguous and simple are relative, context-dependent terms. However, usually algorithms are associated with computers, and when restricted to computers, the term 'algorithm' becomes more precise because 'precise and unambiguous and simple to follow' means 'a computer can do it'. And the term 'computer' is made precise by using the computational notion of a *Turing Machine* (see Turing, 1936, Sipser, 2013, and the Wikipedia entry). The connection with computers is not necessary, however. If a person equipped only with pencil and paper can complete the operations, then the operations constitute an algorithm.

A famous example of an algorithm (dating back at least to Euclid) is finding the greatest common divisor (GCD) of two numbers, m and n.

Step 1. Given two positive integers, set m to be the larger of the two; set n to the smaller of the two.

Step 2. Divide m by n. Save the remainder as r.

Step 3. If $r = 0$, then halt; the GCD is n.

Step 4. Otherwise, set m to the old value of n, and set n to the value of r. Then go to step 2.

(It should be noted that algorithms are by no means restricted to numbers. For example, alphabetizing a list of words is also an algorithm.)

Note: AI and branches of cognitive science are committed to the view that thinking itself is the execution of an algorithm – or perhaps better, the result of many algorithms working simultaneously.

Now for the crucial notion of a *virtual machine*.

Virtual Machines

In common parlance, an algorithm is a recipe for *telling* a computer what to do. But this definition does more harm than good because it subtly reinforces the idea that a homunculus of some sort inside the computer is doing all the work of reading the instructions and doing what they say to do. This is not how computers work. So this folk definition should be avoided when precision is needed, as it is here. Another problem with the folk definition is that it does not do justice the profundity of the notion of an algorithm as a description of a *process*. This move from statics to dynamics is fundamentally important.

Algorithms, when run on an implementing machine, produce *virtual machines*. Such machines instantiate different processes. This is a much better way to think of algorithms, and computation in general.

A virtual machine is a software machine that runs on a lower-level machine, which might itself be virtual, or might be the actual hardware (for a high-level review, see Smith and Nair, 2005). Microsoft Word is an example. Word is a collection of algorithms; but we can think of it here as one large algorithm for making and manipulating texts. While using Word to type a report, say, one is using the Word virtual machine. The Word machine is running on, say, an Apple Macintosh of some sort. To run on the Mac hardware, Word, which is written (currently and mostly) in the programming language C++, has to be turned into binary code that can run on the Mac hardware. This happens (roughly) in two steps: (1) the Word C++ code is compiled into a language called assembler (which is a virtual machine), which is specific to that kind of Mac; (2) assembler is then translated into the binary code which runs natively on the Mac (i.e. on the hardware machine).

Suppose that Lopez, using the Word virtual machine, types 'Teh dog ran home', and then Word automatically fixes the misspelled 'The'. All Lopez has to know is that this is how Word works – that Word will sometimes correct misspellings automatically. She knows also that Word is checking her spelling as she types. She thus knows that the spell-checker (another virtual machine) is running in the background. But she doesn't have to know any technical details about how Word fixed the misspelling, about how Word is implemented, about C++, about the Mac assembler, nor anything about how the Mac hardware works using the binary code version of Word. In fact, Lopez could change to a Dell and still interact with the same Word virtual machine, even though all the implementing details have changed (what the C++ compiles into – Dell assembler – and the native code for the Dell chip architecture).

While she is using the Word machine to write her report, the Mac hardware machine is also working. There are, therefore, at least two substantial machines working (as we noted above, there are really many virtual machines running).

Why talk in terms of virtual machines? Because the *gap* between using Word and the working of the Mac hardware is so large. The gap here is one of knowledge and ontology. As just mentioned, one can be an expert on using Word and yet know nothing of C++ or the Mac hardware Word is running on. But more deeply, becoming an expert on Word requires learning a lot of complicated information that is, in a clear sense, *independent* of the Mac, for as just stated, one could easily transport one's Word knowledge to a Dell computer. As for ontology, the Word virtual machine is a completely different kind of machine than the Mac hardware machine. They do utterly different things – Word checks for misspellings, the hardware, for example, loads information into various registers and accumulators.

Are cars virtual machines – machines for going from A to B? Yes, in a way, but such a move may not be that useful in this case. The gap between being an expert driver and knowing how your car works is large, but at least some knowledge of how your engine burns fuel is usually considered important. And this knowledge starts to bridge the gap between driver and car. What makes the virtual machine idea so important in computers is that the two kinds of machines, Word and Mac, are so radically different from each other. (Note, too, that knowledge about how to be a good driver is mostly kinaesthetic knowledge, whereas knowledge about Word and Macs is mostly declarative knowledge.) Bicycles represent an even closer example. Most cyclists know a fair amount about how their bicycles work (bicycles are far less complicated than computers, and usually, most of the working parts are easily visible). This knowledge of how a cycle works often includes, for example, items like knowing a bit of the physics of maintaining balance. So the gap with riding a bicycle and knowing how one works is quite small. So, the notion of a virtual machine applied to cycles does not increase understanding.

Above, we noted that a computer's complex behaviour requires 'complicated internal states that *are not usefully thought of as physical*'. Virtual machines sideline the physical; the physical becomes a notion that is no longer of prime importance. This is important: it allows virtual machines like Microsoft Word to run on completely different physical hardware, e.g. Macs or Dells.

Consider again the greatest common divisor algorithm from the algorithm part of this section above. Executing this algorithm on a computer or by hand generates a GCD virtual machine. Suppose that we want to find the GCD of 48 and 18. All that matters while using the GCD machine are such facts as these:

> m is first set to 48 and n to 18;
> dividing 48 by 18 gives a remainder of 12;
> so r is set to 12;
> the remainder r is not 0, so now m gets set to 18 and n to 12;
> dividing 18 by 12 gives an r of 6;
> r is still not zero, so m gets set to 12 and n to 6;
> dividing again gives an r of 0;
> the GCD of 48 and 18 is 6.

Note that questions of physical manifestation or implementation are quite beside the point here. Reduction to the physical is no longer all that important. All that matters is that, e.g. m gets set to the bigger number or that the remainder is saved as r. In fact, the GCD virtual machine could be made out of wood or tin cans and rubber bands.

So, the issue that mattered the most to both Behaviourists and Logical Empiricists – the physicality of something – was found to be quite a minor issue after all. It took the invention of the computer for everyone to see this and finally accept it.[3]

3.2 Internal Representations in Virtual Machines

Computer science sowed the seeds of the demise of Behaviourism. Virtual machines allowed computer scientists to introduce and robustly use the notions of an internal state and an internal process. Look again at the three central tenets of Behaviourism, and now consider three *new* tenets:

1 Computer science is *not* the science of computer behaviour. Clearly *inside* a computer, a lot of complex processing is going on.
2 The behaviour of a virtual machine *cannot be* described and explained without make reference to *internal computational processes* (the spell-checking process above, for example).
3 Internal states and processes in a computer cannot be eliminated in any way, nor can they be translated completely into 'behavioural' concepts. (A lot of what a computer does goes on behind the scenes, and the relevant behaviour doesn't need to be apparent to the user.)

The notion of an internal state is just the notion of an *internal representation*. When applied to humans, these internal representations are *mental* representations. And in AI, they are at least quasi-mental representations (see the Third War).

In sum, algorithms for doing things, especially mathematical things, have been with us for millennia (the GCD algorithm above is an example). But once algorithms were implemented on a computer in the mid-twentieth century, the powerful notion of a virtual machine emerged. From here the essential notion of an internal computational representation emerged to give computer scientists truly profound insight into how the mind works: It is a kind of computer. Emphases on physics and physical manifestation became unimportant. Freed from the stultifying need to reduce everything to physics, the leap to AI was obvious and inevitable. So, with Behaviourism dead and mind denied no longer, AI researchers and cognitive scientists set off to understand the human mind, and to build a mind of their own.

But which kind of virtual machine best implements intelligence? That is the matter to which we now turn. Philosophers picked sides and engaged all who disagreed with them.

4

Implementing an Intelligence

1. Four Contenders

Over the history of AI from 1956 (the Dartmouth Conference) to today, one can make out four main software architectures intended by their developers to implement intelligent thought.[1] The four are (in roughly temporal order of their height of enthusiasm):

1 Symbol Processors
2 Connectionist and Artificial Neural Networks
3 Embodied, Situated Cognition
4 Dynamic Systems[2]

It needs to be stressed that each of these four architecture types or *paradigms* comprised a large number of different, individual AI software projects, which differed from each other in large ways. Also, many of these projects lasted years, and even decades, and they were staffed by a large number of AI researchers. Some even produced generations of Ph.Ds. The last three are still pursued today, though in forms modified from the originals. Arguably the first, symbol processors, once the most popular architecture, is fading.

The champions of each of these four included computer scientists, philosophers, psychologists, and other researchers; beginning about 2000, even neuroscientists started to weigh in. Debates between the competing champions formed some of the most intense theorizing and vociferous disagreements in AI. The wars between AI researchers and not-AI-friendly philosophers (such as John Searle and Jerry Fodor) also shaped these debates, though the issues at hand often had more to do with 'aboutness' (see below, Chapter 5) than with architecture per se. Many more pro-AI philosophers championed any architecture that differed from symbol processing (which, not coincidentally, was temporally both the first dominant AI architecture and the first dominant theory of mind in cognitive psychology). For example, Paul and Patricia Churchland championed connectionist networks and neuroscience; Andy Clark championed embodied and situated cognition; and Tim van Gelder championed dynamic systems.[3] As typical in such debates, oversimplification and the construction of straw men were rampant on all sides, making the integration of ideas from different camps into hybrid architectures difficult. As the 'heat' in the architecture wars subsided, comparative discussion has become more fruitful (see e.g. the discussion between Anderson, 2003 and Chrisley, 2003) and hybrid architectures are now commonplace, often under the rubric of 'biologically-inspired' architectures (Goertzel et al., 2010; see also Sun, 2007).

In this chapter, we will discuss these four approaches to architectures for intelligent machines in their listed order, explain their significance, give arguments for and against each architecture, discuss the relevant works of the six philosophers referred to in the previous paragraph along with other philosophers, and explore what is currently the 'winning' architecture.

2. SYMBOL PROCESSORS

2.1 Introduction

The first AI machines, in the 1950s and early 1960s, were *symbol processors*. Symbol processors remained dominant until the sharp rise of connectionist and artificial neural nets. It is impossible to overstate the importance of the move from seeing computers as number crunchers to seeing them as symbol processors. This revelation arguably was the most important insight spawning AI. Indeed, it was more than an insight; it was a revolution, akin to seeing humans as a species of African ape.

2.2 From Number Crunchers to Symbol Processors

Before computers were viewed as symbol processors, computers were viewed as *number crunchers*, i.e. as large numerical calculators. And calculating was all they were used for. ENIAC (Electronic Numerical Integrator and Computer), one of the world's first computers, was intended solely for numerical calculations, even though it was Turing complete. Being Turing complete means that with enough memory, ENIAC could simulate any Turing (computing) machine (see Chapter 1). In practical terms, ENIAC could do anything any modern computer could do, though much more slowly. ENIAC was owned by the U. S. Army, and was operational from 1945 until 1955, spending most of its life at the Aberdeen Proving Grounds in Maryland. It could calculate in seconds what would take a human weeks to calculate. Even so, the idea that a machine like ENIAC could do more than operate on numbers was almost unthinkable. Howard Aiken, who built the first electromechanical computer in the United States, once said 'if it should ever turn out that the basic logics of a machine designed for the numerical solution of differential equations coincide with the logics of a machine intended to make bills for a department store, I would regard this as the most amazing coincidence that I have ever encountered' (1956). Aiken was amazed. But the great discovery that amazed him was that this was no coincidence. (His electromechanical computer was about a thousand times slower than ENIAC, which was entirely electronic. Electromechanical computers were the World War I-era biplanes of computers.) Though Aiken was amazed, he was amazed only over a minor switch from one kind of numerical calculation to another, from differential equations to billing. What would he have thought of the iPhone and the learning machines behind Facebook and other social media sites, all of which run on hardware that adds and subtracts and would have no problem doing department store billing? (Aiken died in 1973.)

There is, however, another error that Aiken made, a deeper one, and this error takes us to the heart of *symbols*. That 'the basic logics of a machine designed for the numerical solution of differential equations coincide with the logics of a machine intended to make bills for a department store' is *not* an amazing coincidence. It is not a *coincidence* at all. In fact, it is a *necessary connection*; an immutable connection, that lies at the heart of computing. How so?

The short answer is that *numbers* <u>are</u> *symbols*. Adding and subtracting, for example, *is* symbol processing. So, an abacus is a symbol processor. Counting on your fingers is processing symbols. But the long form of the answer is what we need here.

The basic logics of computer hardware – *and*, *or*, and *not* gates (for example) – can be arranged to execute any Boolean function (a function that returns true or false given any input of trues and falses).[4] Computers are, really, arrangements of components that compute Boolean functions. Such arrangements of Boolean functions can, in principle, compute anything a Turing machine can compute.[5] So a Boolean machine can compute anything that we regard as computable. That's the deep error Aiken made (although, in his defence, many researchers missed this): A proper arrangement of Boolean functions can do anything we need a computer to do. And only *one* such arrangement is needed. Such arrangements are the basic logics Aiken was referring to. Hence, given one proper arrangement of Boolean components (components executing Boolean functions), we can both provide numerical solutions to differential equations and figure out department store billing – *and* we can manage social media platforms, compute trajectories for distant planetary probes, model climate change, model new medicines' effect on old viral nemeses, analyse SETI signals, send email, publish research, edit pictures and videos, play games, and on and on. In short, with the invention of the ENIAC, *everything* was revealed to be a symbol; it is just that most researchers missed this revelation until the mid-1950s. But finally, at the Dartmouth conference, the real nature of computers was finally perceived: they are general machines for manipulating any collection of symbols.

We note that using the new insight about symbols in computers as the foundation of AI is sometimes called *Good, Old-Fashioned Artificial Intelligence*, or *GOFAI*, for short. As we will see, this term is only sometimes used in a friendly way.

2.3   But What Is a Symbol?

From the previous chapter and the above, we now have the conclusion that *every* computer program ever written is a virtual symbol processor. Furthermore, if we use the standard dictionary definition of the term 'symbol' as *something that stands for something else* we get this same result, because all programs manipulate something that stands for something else – that is, after all, what computing is for. But this hardly seems powerful enough to forge the new discipline of AI in the late 1950s.

The key is that when AI researchers used the term 'symbol', they narrowed it from its dictionary definition. It was used to mean something special. The term meant *language-like* or *linguistic* symbol. Why? Why this *linguistic turn*?[6] What happened was this: When the early AI and computer researchers realized that computers were symbol manipulators, they then quite naturally assimilated computer programs to human languages, which are also symbol systems. Happily, with the demise of Behaviourism (or at least its weakening), along came the early cognitive scientists (starting in the 1950s) who pushed the nascent thesis that humans think in a mental language.[7] (An important book that first explicitly stated this brave new theory was *The Language of Thought* (1975) by Jerry Fodor.) So, with all these ingredients in place, the stage was set for the quest for a machine with human-level intelligence – all that had to be done was to build a *language* machine – a machine that based its 'thinking', its processing, on a language or language-like structure. So, in the early years of AI, the central idea was this:

> Computers are already general symbol manipulators, and human thought seems to be language-based (where the language is not a natural language, but an internal, genetically-supplied language used in the brain), so all we have to do is to program a computer to manipulate language-based symbols and we'll be done.[8]

It was a very seductive idea, this idea that human thought was really thinking in an internal, genetically supplied, brain-based language. This language has not been found to this day; it won't be found (assuming it is there to be found) without a good deal more neuroscience knowledge than we currently possess. The hypothesis that this language exists was officially dubbed the *Language of Thought Hypothesis* (again, first clearly formulated by Fodor). This hypothesis applied only to humans. But AI scientists applied it to computers.

However, though a nice and even useful story, the above is historically inaccurate. The AI scientists were well on their way toward building an intelligent symbol-based computer when Fodor wrote his book. Fodor got the idea not only from the field of cognitive psychology and linguistics in the 1950s and 1960s, but also from AI – a fact that he didn't like (Fodor, 1987, 2000).

The only missing ingredient now was the proper form of manipulation. What could that be? As we will now see, the manipulation of symbols was tied to their specific implementation.

2.4. Knowledge Representation: Frames and Scripts

The whole field of how to represent and process language-like symbols in a computer was (and still is) often called *knowledge representation*. Perhaps the diminution of enthusiasm for classical symbolic processing in knowledge representation is partially explained by the fact that every AI researcher had a different idea of what knowledge was and how it is was to be represented. This rather surprising and somewhat depressing result was established by a well-known survey of AI researchers conducted by Ronald Brachman and Brian Smith and published in 1980. One of their conclusions was:

> Metatheoretic consensus is apparently not one of the field's great achievements
> 1980, p. 5

In short, symbol processing never gave rise to any sort of unified, agreed-upon theory of knowledge representation. And worse, it failed to produce an intelligent computer.[9]

Two classic examples of language-like symbols used as knowledge representations were *frames* and *scripts*. An examination of these will be especially instructive.

Frames and Scripts

Frames and scripts are computer data structures for storing together stereotypical information. (A data structure is any method of organizing data in a computer.) For example, if you go to a public swimming pool in the summer, you expect to see items like chlorinated water, children, lifeguards, and mothers watching their children. If you enter a kitchen, you expect to see things like a stove, a sink, dishes, and knives. If you see a copier in a room, you expect it to have an 'on' switch, to be able to copy writings on papers, to have a place to load paper, etc. If you enter a movie theatre, you expect to scan your e-ticket, get some popcorn, go to your seat, sit through ads, see your movie, and then leave. The idea of a frame or script, then, was to capture these expectations so that a computer or a robot, seeing a copier in the room or going to a movie, would have the same expectations.

Frames were generally used for capturing stereotypical expectations an intelligent computer should have in more or less stable environments – entering an office and seeing a copier, for example. Scripts were generally used for representing stereotypical tasks or procedures – using the copier to copy something, for example, or going to a movie (this was sometimes called 'procedural knowledge'). Frames were first explored by Minsky (1974). Scripts were the invention of Schank and Abelson (1977, though Schank was working on them by 1974).

A note about scripts. Though it is natural now to see scripts as knowledge structures that might help a robot, say, use a copier, Schank's scripts had a narrower focus originally, namely they were used in the research areas called natural language understanding and story understanding. So, the original use for scripts was linguistic, rather than interacting with the external environment.

We will focus on scripts for the remainder of this section. The most famous script was the restaurant script. You enter a restaurant, you order your food, you eat it, you pay for it, you leave. Here is a slightly more detailed example of the restaurant script; this example is still highly simplified.

Script: Restaurant

    Track: Coffee Shop
    Props: Tables, Chairs, Menus, Food, Check, Money
    Roles: Customer, Waiter, Cook, Cashier, Owner
    Scenes: Entering, Ordering, Eating, Exiting
    Entry Condition: Customer is hungry; Customer has money
    Results: Customer is not hungry; Customer has less money;
        Owner has more money
    Scene 1: Entering
      Customer goes into restaurant
      Customer decides where to sit
      Customer goes to selected table
      Customer sits in chair at table
    Scene 2: Ordering
      Customer picks up a menu
      Customer decides on food choice
      Customer orders food from Waiter
      Waiter tells cook the order
      Cook prepares food.
    Scene 3: Eating
      Cook gives food to Waiter
      Waiter gives food to Customer
      Customer eats food
    Scene 4: Exiting
      Waiter writes bill
      Waiter brings bill to customer
      Customer gives tip to Waiter
      Customer goes to Cashier
      Customer gives money to Cashier
      Customer exits restaurant

Figure 4.1 Restaurant Script, adapted from Schank and Abelson, 1977

'Highly simplified' means there is a larger theory behind scripts. For example, Schank assumed, as did many AI researchers, that the mind works like *chemistry*. In chemistry, a relatively small collection of primitive atoms (the 92 naturally occurring elements) combine in complex ways to produce all the molecules we see in our world. Schank's version was that in our minds there is a small collection of primitive concepts that combine to produce all the concepts we use every day. For example, one of Schank's primitives was PTRANS – 'transfer physical location of an object'. So, the real restaurant script, instead of saying 'Customer exits the restaurant' says 'Customer PTRANS to out of restaurant'. This might seem like a small difference, but it is not. The whole theory behind scripts was based on the idea of the existence of *primitive concepts*, for example, concepts we are born with. All of our acquired concepts, like the concept [Computer], are built out of combinations of the primitive ones, so the theory holds.

Just because it works for chemistry, doesn't mean it will work for the mind. The 'primitive concept' theory was actually a hold-over from the Logical Empiricist movement (see the beginning of the previous chapter). It is therefore part of a large collection of simplifying assumptions about the mind (including the one that reasoning is based on logic) that have been proven false; such assumptions have failed to generate any robust theory of mind. The lesson of the past 100 years is that *when it comes to the mind, we should not make any simplifying assumptions.*

2.5. A Philosophical Objection to Frames and Scripts, and to Symbolic Processing in General.

It turns out that Schank and Minsky had matters exactly backwards. Remember that both frames and scripts were designed to handle stereotypical situations. Ask yourself: 'What happens if the situation is not stereotypical?'

Suppose, equipped with its restaurant script, a robot walks into a cafeteria? Now what? What's that queue all about? What are those trays for? Or suppose that our robot walks into a regular restaurant but that someone is pointing a gun at the cashier and robbing the place. Now what?

The key problem is that all the information in a script-based machine is *hand-coded*. This means that a programmer sat down and typed in all the information in the machine's scripts. There is *therefore* no flexibility, no learning, no dynamic fitting of behaviour to environment, no way to handle surprises or anything new. But flexibility, learning, dynamic fitting, and handling surprises are practically the definition of intelligence.

The *hand-coding problem*, as it was called, was an objection pushed by both philosophers and AI cognitive scientists (not all, of course, only some). The hand-coding problem was *the* central problem of the symbolic processing movement for intelligent architectures.[10] The hand-coding problem was overcome by abandoning symbolic processing as the key to AI. Yes, minds are symbol processors, and so are computers, but minds get their symbols from being in the world, from learning from it, and from innate processes (it's not concepts that are innate and primitive, it's processes). Finally, the mind's symbols and their flexibility seem to emerge from lower-level processing. But computers within the symbol processor paradigm got their symbols strictly by hand-coding. This difference made all the difference.

3. CONNECTIONIST AND ARTIFICIAL NEURAL NETWORKS

3.1 Introduction

The essence of connectionist and artificial neural nets (ANNs) is learning and flexibility. In fact, all three alternative (to GOFAI) architecture paradigms – ANNs, embodied cognition, and dynamic systems – enshrined flexibility as a key goal. Now is a good place to point out that the neural network paradigm is the current frontrunner, running far ahead of both embodied cognition and dynamic systems in the quest to build an intelligent computer. Neural networks actually resulted in software, in programs, that work (see below). Both embodied cognition and, especially, dynamic systems seemed to have had their primary impact on psychological theories of the mind, rather than implementations of intelligence. We will discuss embodiment and dynamics in the following sections.

The idea that computing could be done with artificial 'neurons' dates back to McCulloch and Pitts (1943), who developed the first model neuron. The first implemented neural network learning system was the single-layer 'Perceptron' of Rosenblatt (1958). Neural network research thrived as a form of analogue computing in the 1960s. It was eclipsed by GOFAI in the 1970s due in part to Minsky and Papert's highly critical book *Perceptrons* (1969) and in part to advances in digital computing. Research on ANNs was not robustly revived until the mid-1980s, as discussed below.

3.2 How a Neural Net Works: An Overview

Figure 4.2 is a drawing of a very simple ANN, a multi-layer perceptron with one hidden layer.

The input nodes are nodes 1, 2, and 3. The inputs they contain are quantities of some sort ranging (usually) from 0 to 1, inclusive. The middle, or *hidden* layer, is nodes 4 through 9. They get their values from the input nodes; note that every input node sends values to every hidden node, a highly non-biological feature of such networks. The hidden layer is so-called because it is not accessible to the net's surrounding environment; it is affected only by the input layer and affects only the output layer. The output layer is one node, node 10. It gets its value from the nodes in the hidden layer. The output node is how the network informs something in its environment,

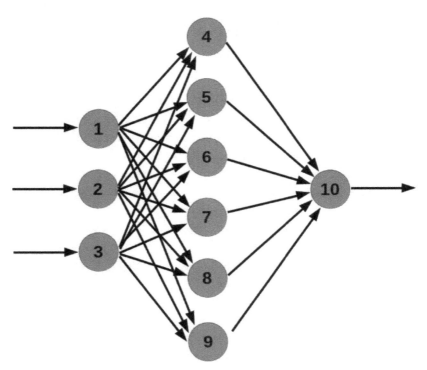

Figure 4.2 A simple feedforward ANN. Information flow is from left to right.

usually another network or an observer of some sort, what it did with the input values. These days, neural nets use many hidden layers and the total number of nodes is usually quite large. Usually the values contained in each node are called that node's *activation* values or levels.

The arrows (often called arcs) in our network are transmission lines. They are for transmitting the activations of the input layer to the hidden layer, and from the hidden layer to the output layer (node 10). But the arrows are weighted. That is, each arrow is fixed with a weight, usually between 0 and 1. This weight determines how much of the signal gets through to the next layer. This is accomplished by multiplying the weight by the activation value of the node in the layer to the left. Our network is *fully connected*, meaning that each node in layer L is connected to *every* node in layer $L + 1$ (terminating in the output layer).

The input nodes are capable of activation, memory and the ability to transmit their activations. The nodes of the hidden layer are little computers: they compute that node's activation level. For each node of the hidden layer, first, the node computes the sum of each input multiplied by the weight on the arc that input traversed; then, if that sum is greater than some fixed threshold, that node fires, transmitting its value to the next layer across the arcs leaving that node, in our case, the arcs to the output node, node 10. More mathematically, for the nodes in the hidden layer, we compute a sum of products:

The activation level of node 4 =
    (Activation level of Node 1 * Weight on arc from node 1 to 4)
    +
    (Activation level of Node 2 * Weight on arc from node 2 to 4)
    +
    (Activation level of Node 3 * Weight on arc from node 3 to 4)
The activation level of node 5 =
    (Activation level of Node 1 * Weight on arc from node 1 to 5)
    +
    (Activation level of Node 2 * Weight on arc from node 2 to 5)
    +
    (Activation level of Node 3 * Weight on arc from node 3 to 5)
    . . . And so on for nodes 6, 7, 8, 9.

The activation level for node 10 is computed in the same way, but using the activation levels and arc weights from the nodes of the hidden layer.

All of this is captured in the following algorithm. (This algorithm is written in what is called *pseudo-code*: an English version of an algorithm and its mathematics, written to capture the essence of the algorithm. As is common, the ';' indicates a comment, which is computationally inert. And * indicates multiplication.)

```
; Set up the initial definitions
1. Definitions
    1.1. Let the weight on each arc be designated Wij, where node i is in the
         previous layer and where node j is the node under focus, i.e. the one where
         the arc terminates
    1.2. Let Vx be the activation value of node x.
; Set up the input layer with its initial values
2. For i from 1 to 3 do:
    2.1. Set Vi to its input activation level
3. End Do
; Compute the activation values for the nodes in the hidden layer
4. For j from 4 to 9 do:
    4.1. Set Vj to 0
    4.2. For i from 1 to 3 do:
        4.2.1. Set Vj to Vj + (Vi * Wij)
    4.3. End Do
5. End Do
; Compute activation value of output layer
6. Set j = 10
    6.1. Set Vj to 0
    6.2. For i from 4 to 9 do:
        6.2.1. Set Vj to Vj + (Vi * Wij)
    6.3. End Do
7. End Program
```

Figure 4.3 The pseudo-code for feedforward part of the artificial neural network in Figure 4.2. This algorithm computes the left to right activation levels for all of the net's nodes.

Learning

We haven't yet discussed the most important fact about ANNs, which is: *they learn*. All kinds of machine learning techniques exist. Machine learning is a large field of study, dating back to the 1960s. Two properties make the learning of multi-layer networks, like the kind in Figure 4.2, important. Property 1 is that information represented is *distributed* over the nodes at each layer. For example, getting a neural network to learn to pick out dogs in pictures of dogs and cats requires first distributing the information in such pictures over the input layer. This distribution property helps make neural nets robust in the face of damage. For example, if several nodes stop functioning after the dog-recognizing network has learned to recognize dogs, the network will still recognize dogs. Also, distribution helps the network recognize occluded dogs, dogs that are partially hidden by cats, say. Property 2 is that ANNs capable of nonlinear classification are *multi-layered* (single-layer perceptrons can learn linear classification tasks). Multi-layer networks can learn important abstractions that are only implicit in their input data. Other machine architectures have a hard time with this problem.

Getting neural networks to learn requires changing weights across *all* the layers. Neural net learning is adjusting the weights so that the output layer exhibits the right answer. So, assume the weights are set randomly. Then the output layer will very likely get the answer wrong. The object now is to adjust the weights slowly throughout the network so that the network eventually computes the right answer, which it exhibits in the output layer . . . But how to change all the weights so that the network produces the right answer eluded researchers.

The first major breakthrough in getting nets to learn was the application of a mathematical discovery often associated with Seppo Linnainmaa, a Finnish mathematician (for details, see e.g. his 1976). But many others had similar ideas. Condensing a lot of mathematical history, all of this work culminated in the *backpropagation algorithm*. Basically, this algorithm adjusts weights by propagating errors at the output level backwards through the network with each arc getting assigned some of the 'blame' for the error. Adjust the weights accordingly, and the error will diminish. This algorithm eventually gets the weights right, and the network produces the right answer. Learning, then, is adjusting weights by apportioning blame.

(The story of neural nets that learn is so large that we have to leave out much technical information and interesting history. Just the mathematics involves multidimensional vector algebra, calculus, differential equations, and geometry. The reader is encouraged to start with the central *Parallel Distributed Processing* volumes, one by Rumelhart and McClelland, and the other by McClelland and Rumelhart (both 1986). These volumes ushered in the paradigm change from symbolic AI, discussed above, to what was called at the time *subsymbolic AI*. From here, the literature grows quite large, but is easily obtained; the advances have been stunning – see section 3.4.)

3.3 Philosophy Meets Neural Networks

Artificial neural networks profoundly altered the philosophical landscape in the philosophy of mind and the philosophy of cognitive science. Some philosophers saw ANNs as a paradigm shift on the level of the General Theory of Relativity: Finally, a genuine theory of mind was in the offing. And so, therefore, was the path to a genuinely intelligent machine. Paul and Patricia Churchland were leaders of this new army of hope. With both mind and machine poised to be reconceived, the future did indeed look bright. In his book for nonspecialist readers (*The Engine of Reason, the Seat of the Soul*), Paul Churchland says:

> Fortunately, recent research into neural networks, both in animals and in artificial models, has produced the beginnings of a real understanding of how the biological brain works – a real understanding, that is, of how *you* work, and everyone else like you ... [W]e are now in a position to explain how our vivid sensory experience arises in the sensory cortex of our brains ... We now have the resources to explain how the motor cortex, the cerebellum, and the spinal cord conduct an orchestra of muscles to perform the cheetah's dash, the falcon's strike, or the ballerina's dying swan ... [W]e can now understand how the infant brain slowly develops a framework of concepts with which to comprehend the world ...
>
> Churchland, 1995, pp. 3–4

Not everyone was hopeful. Jerry Fodor, for example, took a dim view of the matter. He thought it was all hype. Fodor was a committed fan of symbol processing. In a critical review of the book from which the above Churchland quote is taken, Fodor said of the idea behind neural networks:

> ... this is just an unedifying kind of fooling around.
>
> Fodor, 1998, p. 88

Discussing the fact that the layers of ANNs can be thought of as defining a multidimensional space (such a move is useful for developing the mathematics of ANNs), Fodor said:

> Saying [as Churchland does] that we conceptualize things as regions in a multidimensional space is saying no more than that we conceptualize things. The appearance of progress is merely rhetorical. If you want vectors and dimensions to be a breakthrough and not just loose talk, you've got to provide some serious proposals about *what the dimensions of cognitive space are*. This is exactly equivalent to asking for a revealing taxonomy of the kinds of concepts that people can have. Short of puzzles about consciousness, there isn't any problem about the mind that's harder or less well understood. It goes without saying (or anyhow it should go without saying) that Churchland doesn't know how to provide such a taxonomy. Nor do I. Nor does anybody else.

Fodor, 1998, pp. 88–9. For much more technical and detailed criticisms of neural nets, see Fodor and Pylyshyn, 1988, and Fodor and McLaughlin, 1990

More reserved philosophers who were nevertheless still impressed said things like:

> In the space of about a decade, connectionist approaches to learning, retrieval, and representation have transformed the practice of cognitive science . . . As a philosopher and cognitive scientist long interested in the connectionist program, I remain convinced that the change is indeed a deep and important one. This conviction makes me, I suppose, a neural romantic.

Clark, 1993, p. ix

(Another very good, measured discussion of the philosophy and philosophical objections behind the rise of ANNs, including a good technical introduction to networks, is Bechtel and Abrahamsen, 1991. And for a good general introduction, see Garson, 2018.)

Our view is that Fodor made some very good points. Neural nets have proven *not* to be the royal road to a robust theory of mind and brain (at least so far). Nor have they led to building human-level artificial intelligence. Neural nets are bad, for example, at making inferences, which is arguably the essence of intelligence. But Fodor clearly underestimated how useful large ANNs could be: such machines form the backbone of much of our modern world (see the next section). Fodor underestimated the practical side of ANNs. That much is understandable. Everyone underestimated automobiles, aeroplanes, and computers, to name three. But the excitement over ANNs is also understandable, and to be applauded. We now know vastly more than we did, thanks to the enthusiasm of the neural romantics.

3.4 In Sum

It is impossible to overstate the importance in AI, especially in practical or applied AI, of the numerical turn ushered in by the discovery of a robust learning algorithm for ANNs. First and foremost, they made possible the discovery that large ANNs with many hidden layers can learn very sophisticated skills. (Such networks are called *deep learning networks* – there are many types and classes of such networks; see LeCun, Bengio and Hinton, 2015) Here is a small list of the accomplishments of deep learning networks:

1 The marketing systems of Amazon, Facebook, and Google (etc.); these systems learn their customers' preferences.

2 AlphaZero, the reigning chess and go champion of the world (as of 2019); AlphaZero can beat any human at chess

3 Watson, a computer built by IBM, is the reigning world champion *Jeopardy!* player; but IBM is deploying Watson for serious work, such as helping doctors with diagnosing illness.

4 DeepStack, a world-class champion at playing heads-up no-limit Texas hold'em has beaten several professional players.

(For more, see Supplement 1 to the Third War, where we argue that, as sophisticated as these machines are, they don't advance us toward the goal of building machines with human-level intelligence.)

In fact, AI's current extraordinary relevancy to our lives is in large part due to ANNs. It is worth saying again that (1) to (4) are skills that the relevant machines *learned*. The machines were not hand-coded to play, e.g. Texas hold'em. It is probably impossible to build a script-based or any other symbol processor machine to play world-class Texas hold'em.

But the relevance of ANNs to psychology and to the goal of understanding the human mind is much less clear. One could even argue that ANNs' contribution to the sciences of the mind is minimal. Most of the mind and its brain remain *terra incognita*. This is easily explained. The information flow was from neuroscience to ANNs. The cognitive scientists and AI researchers borrowed from neuroscience. But the reverse never seemed to happen, or never seemed to happen with any robustness. This is mostly due to the simplicity of the artificial 'neurons' that such networks employ, which have evolved only minimally since the models of McCulloch and Pitts. Such models vastly under-represent the complexity of real neurons (see e.g. Gidon et al., 2020, for a demonstration that dendrites in single neurons can perform 'xor'). Hence ANNs, even with deep learning, vastly under-represent the complexity, and the capability, of biological neural networks.

Finally, have ANNs, as advertised, really solved the hand-coding problem? Have they achieved the long-sought flexibility indicative of intelligent minds? The answer is, 'Somewhat.' Yes, deep learners are far more flexible than hand-coded AI systems. And they can learn to do amazing things. But as noted above, and discussed more thoroughly in the Third War, deep learners are useful, but not really intelligent. Somehow humans are deep learners about everything, and they possess this capability using a machine that weighs only about 1361 grams (3 pounds). How this could be remains a mystery.

Segue: Attacks on the AI Hypothesis.

As we have seen (and will see more of), AI has been under attack almost from its beginning. Philosophers led the charge (e.g. John Lucas, 1961 – see Chapter 1), but they were joined by computer scientists (e.g. Weizenbaum, 1976), mathematicians (e.g. Penrose, 1989), and psychologists (e.g. Spivey, 2007, see section 4, below). Of course, not everyone was in attack mode. But many were. The very idea that human brains are computers and humans are robots was, to many, simply and obviously false.[11] Consequently, the attacks tended to try find something special that human thinking had which computation lacked. Cognitive scientists and machine intelligence researchers tended to focus on two interesting differences: humans rely heavily on their environments for their thinking, and human neural processing is dynamic, like a moving fluid. Approaches to AI architecture that took one or the other of these features of human cognition as their foci are called *embodied (situated) cognition* and *dynamical systems*, respectively (see next). We will return to this 'something special' analysis when we consider two other proposed special properties in Chapters 5 (semantics: humans think about things, but computations are not about things) and 6 (humans but not machines have the ability to solve the frame problem). Both of these latter 'special properties' were pursued almost exclusively by philosophers.

4. EMBODIED, SITUATED COGNITION AND DYNAMIC SYSTEMS

4.1 Introduction

We treat together these two approaches to the mind because they share a fundamental property: they both deny what the symbol-processing camp asserts: thinking is manipulating language-like or language-based mental symbols of the kind discussed above, in section 2.3. In fact, some researchers favouring embodiment and dynamics deny that the mind contains any symbols whatsoever; some even deny that that the mind contains any *representations* at all, whether symbol-like or not. And then there are those who go even further: denying that an intelligent computer is even possible, then asserting that a human-made intelligence will have to be a robot of some sort whose 'thoughts' *emerge* from low-level, continuous, dynamical, neural-like systems.

4.2 Embodied Cognition

Embodied cognition (EC), sometimes also called 'behaviour-based robotics', is an approach to AI that takes the opposite approach to minds that the symbol-processing camp does. First, EC is primarily about robotics: the intelligence of the machine is devoted to interacting with its environment through its body. Instead of a frame-based system that reasons about restaurant behaviour, an EC robot would actually go to a restaurant and order food directly (of course, it wouldn't eat the food, being a robot, but it might subject said food to various chemical analyses). Second, EC privileges perception and action, and builds up intelligent behaviour out of those two. Third, more complex behaviours can emerge by adding to the initial low-level perception and action more and more layers of sensors (which, in the higher layers, sense the outputs of other, lower-level sensors) and more and more layers of action-producers. Such layered architectures are called *subsumption architectures* (Brooks, 1999).

The EC movement developed from three historical currents. The first, and most significant for AI, was the search for an architecture that allowed autonomous and robust motility for relatively simple, inexpensive robots. Rodney Brooks, a pioneer in this area, showed that distributed control systems, one for each limb, would allow a multi-legged 'insect-like' robot to cross obstacle-laden terrain (Brooks, 1989). Centralized symbol processing was not required; local (to the limb) sensitivity to environmental conditions was (Brooks, 1991). The second contributor to EC was the revival, within psychology, of the Behaviourist focus on perception – action loops as the foundation of intelligent behaviour, particularly in 'lower' animals. J. J. Gibson, one of the primary advocates of this approach, argued that intelligent behaviour involved no 'information processing' at all (Gibson, 1979), an argument taken up by more recent 'radical embodied' psychologists such as Anthony Chemero (2011). The final thread was philosophical. It combined the general distaste for GOFAI held by critics such as Dreyfus (1972) with worries on the part of symbolically minded philosophers that 'lower' organisms do not employ symbol processing (Fodor, 1986).

It is easiest to understand EC by contrast with GOFAI. Here is a high-level sketch of the classical, symbol-based view of the mind.

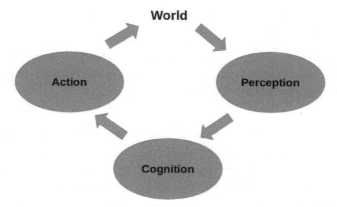

Figure 4.4 Derived from Brooks, 1999, p. viii. Note the linearity of the information flow: from the world, to perceptual processors, to cognition (i.e. thinking), to action (limb movement, for example). There is no immediate feedback from, say, Cognition to Perception. Perception, Cognition, and Action constitute the *mind*. This model is what the ECers deny.

Opposing Figure 4.4, Brooks proposed this:

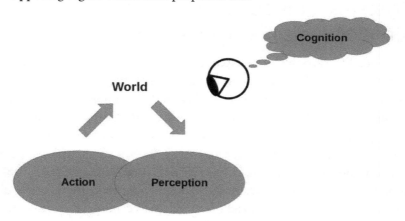

Figure 4.5 Derived from Brooks, 1999, p. xi. Note that action and perception intersect, giving a kind of 'perceiving-is-doing/doing-is-perceiving' view of the mind. Also note that cognition is absent. It is crucial to see that for Brooks and other ECers, cognition is only in the eye of the beholder – hence, the eye and the imagination that 'cognition' is occurring. A behaving being, e.g. a creature, robot or human only seems to be cognitive, to be cogitating, due to its action-based perceptions and perception-based actions. Cognition, in a real sense, according to Brooks, is not real; it merely appears real to an outside observer. This is like sunrise and sunset here on Earth. The sun seems to rise in the morning, but it doesn't really rise; the Earth rotates relative to the sun.

To get a further sense of how different the embodied cognition research project is, here is Brooks:

> I believe that mobility, acute vision, and the ability to carry out survival-related tasks in a dynamic environment provide a necessary basis for the development of true intelligence.
>
> Brooks, 1999, p. 81

This focus on intelligence as an ability to *explore* an environment using active senses and directed behaviour, and to learn from such exploration even as the environment unpredictably changes, is perhaps the most significant outcome of the EC movement. Following the introduction of flexible robotic platforms such as the iCub, it has led to the emergence of developmental robotics as a robust area of research combining insights from developmental psychology with EC-influenced AI (for a comprehensive review, see Cangelosi and Schlesinger, 2015).

4.3 From a Three-way Duel to Hybrid Systems: Embodied Cognition + Connectionism + Symbol Processing.

Are the ECers right? Are psychologists wrong that there are mental representations like concepts, word meanings, ideas, propositions (mental sentences), and reasonings about all of these? For example, are psychologists wrong that we thinkers constantly use *concepts* for understanding and dealing with everything we encounter every second of every day (Murphy, 2002)? Concepts are the Platonic ideal of symbols, if anything is. Do we not use concepts to understand and deal with dogs, cats, bears, octopuses, robins, eagles, forks, knives, people, trees, children, sun, rain, snow, wind, dirt, viruses, cars, clothes, and on and on ... ? Perhaps concepts do supervene on lower level sensorimotor computational processes, but that doesn't mean there aren't concepts. Dogs, after all, supervene on lower-level biochemical processes like respiration and digestion and cell-growth, but that doesn't mean there aren't dogs. Dogs exist. And we can think about them. Any approach to AI that denies this is probably in big trouble.

However, building robots using scripts and frames was very clearly not working. And it was not going to work, due to the hand-coding problem (see section 2.5, above). Yet there clearly are concepts and script-like representations for many things, including restaurants. Most importantly, the cognitive work we do with concepts clearly exists, namely *inference*. Being able to draw inferences is crucial to nearly all animals. We infer information from what we perceive, what we remember, what we predict, what we conceive, and what we experience. Unfortunately, ANNs do not draw explicit inferences, and neither do embodied cognitive systems. (Neural nets can learn inferences, but this is different from constructing them in real time as needed.)

In sum, we now know several, related, but contradictory things:

1 Humans have and use concepts, and concepts seem a lot like symbols,
2 Symbols are essential for inferencing, and inferencing is essential for cognition,
3 Neural nets learn complicated, abstract information,
4 If neural nets represent symbols at all (a matter of debate) they represent them as vectors composed of numerical values (see section 3.1),
5 Embodied cognition robots do work reasonably well, especially on narrow, fixed tasks, like roaming around a home vacuuming.

We wind up concluding that to build an intelligent machine, we have to have symbols, but we cannot use symbols and must instead use embodied ANNs. This is an intolerable situation. Worse, every AI researcher believed that he or she was following closely the architecture of the mind. Minds do appear to use symbols, especially concepts and even rules as in the restaurant script, but minds also learn by adjusting synaptic weights across very large connections of synapses, and the mind-plus-its-body appears to have components resembling some sort of subsumption structure.

Perhaps the key is the use of *hybrid systems*. Suppose we divided up the cognitive jobs. Brooks' perception and action could be some combination of many neural nets nested into a subsumption architecture (see section 4.1). And all of that information could be shared with some sort of symbol processor. Such hybrid systems exist and are relatively successful. Most, however, only combine artificial neural nets with symbols, omitting any subsumption architectures. One such system is the program Clarion (see Sun, 2018, 2013, Sun and Zhang, 2006, and Sun et al., 2005). Though Clarion does not use a subsumption architecture, it does combine low-level implicit information processing with high-level explicit symbol processing. Another net/symbol hybrid is Hummel's and Holyoak's experiments combining the distributed representations of connectionism with the structure of higher-level symbols (1997). Still another is LIDA, an implementation of the 'global neuronal workspace' (GNW) theory developed by neuroscientists (Franklin et al., 2014; see Chapter 7 for more on the GNW).

But are the hybrids walking among us? It is obvious that we do not have robots walking among us with human-level intelligence. But this, ultimately, is precisely what hybrids are designed for, at least in the far term. So, we can infer that, so far, the hybrid approach to human-level AI has not worked (though it has worked for developing useful cognitive models of small portions of human perception and cognition). The quest, then, continues for an intelligent machine that is a human peer.

4.4 Dynamic Systems Change the Quest

The advocates of a dynamic systems approach wanted to change the quest completely (for three central texts, see Chemero, 2009; Port and van Gelder, 1995; and Spivey, 2007). The proponents of this approach hold that the computational paradigm which dominates cognitive science is wrong. They follow Gibson (1979) in claiming that thinking is *not* computing, period. Thinking is an emergent property of the dynamic system we call the brain. The dynamic systems proponents make two claims: (1) Developing a useful psychological theory of minds requires thinking of them as dynamical systems; and (2) Any hope of building a human-made intelligence requires building a thinking dynamic system.

What are dynamic systems? And what is their role in cognitive science?

Let's begin answering these with van Gelder's (1998) answer to the second of the above two questions. Van Gelder claims that cognitive systems (e.g. humans) are dynamical systems, and that the study of cognition ought to be couched in the language of dynamical systems. Dynamic systems make up a large category. The claim is that within this category, one will find a specific class of dynamic systems that think; van Gelder calls this the *dynamical hypothesis*. (As is usually done, we will treat 'dynamic system' and 'dynamical system' as synonyms.) Note, the dynamical hypothesis, like the computational hypothesis with which it contrasts, is really two claims: an ontological claim about the nature of cognitive systems, and an epistemological (philosophy of science) claim about the science of cognitive systems, i.e. about how best to study and explain cognitive systems.

Now for the first question: What's a dynamic system? Let's begin with the notion of a *system*. A system is any collection of physical objects with various properties that interact with each other. Any system whatsoever can be described by a set of interdependent variables. What distinguishes a *dynamical* system from a computational system is what these variables range over. In a computational system, the variables – the data structures – can range over anything: numbers, people, restaurants, bank balances, cake ingredients, etc. But in a dynamical system, the variables range over numbers, and only numbers. These numbers denote *quantities* that measure certain *time-dependent* properties of the physical objects which make up the system in question. Of central importance in any description of a dynamical system is the notion of the *rate of change* of its relevant quantities. This is what it means to be interested only in time-dependent properties. And it is why all descriptions of (continuous) dynamical systems involve *differential equations* of one sort or another. Examples of time-dependent properties of a system are temperature, velocity, chemical density, firing rate, and recovery rate. All of the measured quantities of a given system considered together define what is called a *state-space* (sometimes called a phase space). At any given time, the system in question is completely defined by its position in this abstract space. The behaviour of the system over time is, from the dynamic systems' perspective, just its path through this space. Differential equations describe this path.

An important assumption of dynamic systems is that the rates of change of the relevant quantities are continuous (in the mathematical sense) and not discrete. Interestingly, rates of change need not actually be continuous, but the supporters of the dynamical hypothesis take it as a given that the best way to describe such rates of change is as continuous processes. The reason for this is that dynamical system theories are tied necessarily to *time*, and time is usually regarded as continuous (see Port and van Gelder, 1995). For example, it is implausible that the processes that constitute neural firing patterns in brains are genuinely continuous (sodium and potassium ions, key ingredients in transmitting neural singles across synapses, are discrete entities, after all, as are all ions), but it is sometimes useful to assume that such processes are best *explained* as continuous processes.

The assumption about continuity and its utility in explaining cognition puts the dynamical hypothesis squarely at odds with the computational hypothesis, for the computational hypothesis holds that cognition and perception are (or are best described as) the execution of algorithms over discrete mental representations, whether they be symbolic, connectionist, or tied to embodiment. It has even been suggested that continuous systems can compute a larger class of functions than discrete systems, e.g. by Smolensky (1988, p. 18): 'if there is any new theory of computation implicit in the subsymbolic approach [to cognitive science], it is likely to be a result of a fundamentally different, continuous formulation of computation.' Such suggestions ignore the fact that the states of any computer must be *observable* for its behaviour to count as computation. The observable states of any dynamical system are both discrete and finite, like those of a Turing machine (Fields, 1989; for the relevance of this point to quantum computing, see Supplement 4.1 below).

A second important property of dynamical systems is their reliance on tight feedback loops to achieve equilibrium with their environment. Such equilibration is often called 'self-organization'. Proponents of the dynamic systems approach frequently point to Watt's steam engine governor as an example of such equilibration (see Figure 4.6; for the use of the governor in attacking the computational + representational hypothesis, see van Gelder, 1995).

The steam engine governor is a device designed to keep steam engines from exploding. The governor is fixed to a steam pipe. The governor consists of two balls mounted on arms attached to, and spinning around, a spindle. As the engine runs faster, the spindle spins faster causing the balls on the either side of it to rise due to centrifugal force. The rising balls cause a valve to close because of mechanical connections between the arms and the valve. The restricted valve decreases the amount of steam flowing, and hence the pressure of the steam, which in turn causes the engine to slow down, which in turn causes the governor to spin more slowly. The slower spin causes the balls on the arms to drop, opening the valve, which causes more steam to flow, which increases the pressure, which causes the engine to accelerate, etc. In this way, a relatively constant pressure inside the engine can be maintained. This equilibration can be described as a path in a dynamic systems phase space, usually as a cycle

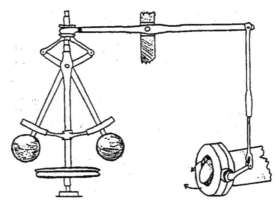

Figure 4.6 Watt's steam engine governor. From J. Farey, *A Treatise on the Steam Engine: Historical, Practical, and Descriptive* (London: Longman, Rees, Orme, Brown, and Green, 1827)

around an attractor. The Watt governor is touted as a correct analogy for cognitive processing; as van Gelder said: 'the Watt governor is preferable to the Turing machine as a landmark for models of cognition' (1995, p. 381).

The key to seeing the Watt governor as the landmark model for thinking dynamic systems is van Gelder's claim that the governor does what it does using no representations whatsoever (1995). He argues at length that arm angles (the obvious candidates for the relevant representations of steam pressure and engine speed) are not representations at all. However, this claim – that the arm angles do *not* constitute representations – has been refuted by Bechtel, 1998. It is now clear that the arm angles *are* representations – continuous representations, but genuine representations nevertheless.[12]

The final and crucial property of dynamic systems is that they give rise to *emergent behaviour*. Humans' and other animals' behaviour emerges from the dynamical systems called brains-in-bodies, according to the dynamic systems proponents. And, on the implementation side, what the dynamic systems proponents were (are) hoping for is that given the right materials, the right differential equations, various meanderings through various phase spaces, and subtle, fluidic interactions, cognition would somehow emerge in all its nonalgorithmic glory. Such emergent cognition and concomitant behaviour seem to be the proverbial free lunch.

There are today good computer models of many aspects of cognition: learning, speech, analogy-making, language understanding, walking, inferring, reasoning, and so forth. All of these models are partial; we clearly have a long way to go before we can stroll through a park having a conversation with a robot. But there is not a single implemented, working model of any aspect of cognition from the dynamic systems camp. We suspect this is no accident. How would you build one? If the dynamic systems proponents were to ever build an intelligent artefact, what would it look like? One can imagine only that it would be something like an aquarium filled with different coloured oils of different weights and densities heated to different temperatures all swirling about and intermingling with each other and giving rise to beautiful coloured, fractal paisleys. And out of this beautiful, chaotic interaction would emerge ... what? ... Wonderful cognition? This seems very unlikely, to put it mildly.

4.5 Conclusion

This completes our discussion of the four most important architectures deployed in the quest to build a machine with human-level intelligence.

Pretty obviously, none of them have worked, if the goal is to produce human-level machine intelligence. Of course, ANNs work very well if the goal is to build useful machines that can help us with our jobs and tasks. And there are robots, of a sort that requires some human intervention, everywhere, from our medical operating rooms to Mars. But nowhere is there a machine even mildly competent at English or German or Mandarin. And self-driving cars have already killed people.[13]

Second War Supplement 4.1: Quantum Architectures

The idea that quantum systems could function as computers was introduced by Feynman (1982). Quantum computers differ from classical computers in two fundamental ways. First, every operation performed is unitary, and hence strictly conserves information and is fully reversible; quantum computing can, therefore, be viewed as a generalization of reversible classical computing. Second, while the final state of a quantum computation is well-defined theoretically as a unitary transformation of the initial state, the final state cannot, in general, be unambiguously determined. The final state of the computation must instead be characterized by (typically multiple) measurements for which probability distributions of possible outcomes can be (classically) calculated.

Deutsch (1985) initiated the formal theory of quantum computing by providing a specification of a universal programmable quantum computer (a 'quantum Turing machine'). Shor (1994) confirmed the expectation that quantum computers could outperform classical computers on practical problems by demonstrating a quantum algorithm to factor large numbers exponentially faster than any known classical algorithm. Lloyd (1996) showed that Feynman's (1982) suggestion that a quantum computer could simulate the dynamics of any (closed) quantum system is correct. The theory has since developed rapidly; for a standard textbook presentation, see Nielsen and Chuang (2000). The main current application areas for quantum computing are cryptography and machine learning, though in both cases hardware development lags substantially behind algorithm development. Apart from machine learning, quantum computing has had negligible impact on AI; see Ying (2010) for a review and forward-looking discussion.

Despite its infancy during the AI Wars, quantum computing made one substantial contribution to the philosophical attack on AI: it provided the theoretical backdrop to Roger Penrose's (1989) reformulation of Lucas' (1961) argument that no classical process could explain human creativity, and hence human intelligence (see Chapter 1). As did Lucas, Penrose focused explicitly on consciousness as an explanandum; Penrose (1989) can be regarded as reintroducing consciousness as a battleground in the AI Wars, and hence as setting the stage for the expanded discussion of consciousness since 2000 (see Chapters 5 and 7).

The association of consciousness with quantum theory has a long history, with von Neumann (1958), Wigner (1963), Orlov (1982), and more recently Schwartz, Stapp, and Beauregard (2005) arguing explicitly that the observer's conscious perception causes the 'collapse' of a quantum state. Penrose reversed this association, arguing that the collapse of a particular kind of quantum state, a superposition between alternative Planck-scale spacetime geometries, causes (or ontically, is) consciousness. This proposal implies, clearly, a deep panpsychism (or 'panexperientialism') in which all parts of the physical universe are conscious all of the time. This kind of universal panpsychism appears radical from an AI Wars perspective, but is surprisingly common among physicists, especially in the form of Wheeler's (1983) proposal that 'observer-participancy' is the sole ontic fundamental in cosmology.

The Third War:
Mental Semantics and
Mental Symbols

5

The Strange Case of the Missing Meaning: Can Computers Think About Things?

1. Introduction: Semantic Meaning and Intelligence

This chapter is about meaning – the meanings of sentences, of actions, of thoughts, of computations. Meaning was one of the main battlegrounds in the AI Wars. Philosophers attacked the AI project by pointing out that while humans and human lives are rife with meaning (reading and understanding this meaningful sentence, for example), computers seem to be without any meaning whatsoever. Computer programs and their data structures seemingly mean nothing at all to the computers running them (though, of course, they mean something to us, their builders). The importance of this difference between humans and machines cannot be overstated. (1) Humans but not computers have meaningful internal processes (neural processes in humans, electronic processes in computers). (2) Humans but not computers are intelligent (as of this writing). So, (3), perhaps (1) explains (2) – computers are boxes of meaningless electronics, whereas humans are soft squishy bags of meaningful wetware. This chapter explores this idea and the associated philosophical attacks on AI.

2. The Problem of Meaning

The realm to be explored here is merely the latest incarnation of the *problem of meaning*: *How do sentences, actions, deeds, and thoughts have meaning?*

What is meaning? How are we able to grasp, share, and manipulate meaning? Humans, philosophers mostly, have been working on this problem for a very long time. Plato, in his profound dialogue, *Euthyphro*, shows Socrates and Euthyphro struggling to grasp the meaning of the word 'piety' (something like 'dutiful virtuousness', especially, religious duty). Here's the dialogue's structure. Euthyphro and Socrates accidentally meet at a trial court. Euthyphro insists that he is fully in command of the notion of piety and knows what is pious and what is not. Socrates sorely needs to understand this notion and the meaning of the word 'piety' because the powerful elite of Athens have charged him with being impious – a serious accusation. Naturally, Socrates is thrilled to have happened on Euthyphro, for the latter can now enlighten him. Euthyphro proposes a definition. Socrates rejects it, pointing out why it won't work. Euthyphro agrees. Euthyphro and Socrates cycle this loop four more times. In the end, Euthyphro and Socrates come up empty-handed: 'piety' remains undefined. Though they know how and when to use the word, they cannot say what its meaning is. Yes, they can roughly characterize the meaning (as we have done here). But beyond that, there is only silence.

Generalizing: saying what the meaning of a word is turns out to be very difficult. But why should this be so?

One reason is that the problem of meaning is in fact a version of one of the deepest problems in philosophy. And this latter problem captured the imaginations of philosophers long before Plato.

The Milesians (from in and around the ancient city of Miletus, western coast of modern-day Turkey, circa the sixth century BCE) are among the first philosophers we have any serious record of. One central question of this group of philosophers was '*Is there an unchanging reality behind the changing appearances of things?*' Let's call this the *Milesian Problem*. This is clearly a deep and difficult problem. Most philosophers assume the answer is *yes*, but to this day no one has offered a successful, in the sense of broadly accepted, theory of what that reality is. Nor has anyone been able to say, with any definitiveness, whether there is more than one such unchanging reality behind things. To see the Milesian problem's depth and difficulty, note that the question about unchanging reality is also *science's* question. Obviously in one sense, science has done a great job in answering it. The answer is the very science we all use every day – this is the reality behind things (see e.g. Sellars' (1962) contrast between the 'manifest image' of ordinary experience and the 'scientific image'). The puzzle, though, is that the entire history of science can be seen as simply unearthing more things with changing appearances (quarks, atoms, genes, viruses, cells, organisms, species, galaxies, galactic superclusters, gravity waves, the universe). It is far from clear, therefore, that science is revealing any *unchanging* reality *behind* the changing appearances of things. Rather, science is unearthing more and more superficial things with changing appearances. Any unchanging reality behind these superficial things continues to elude us. From this perspective, the Milesian problem appears intractable.[1]

To see that the problem meaning is very similar to the Milesian problem of whether or not there is an unchanging reality behind things we need only note that a set of well-known words with somewhat fixed and relatively unchanging meanings are constantly used in new ways to deal with an open-ended variety of situations. The result is that sentences and their meanings are (usually) new, but the meanings of their constituent words are not. Of course, meanings of words do change, and new words are added to languages. But all this change happens quite slowly compared to the second-by-second production of new sentences with new meanings (such as this one).

The human ability to produce new sentences with new meanings using old words is so robust that it is hard to imagine that we moderns are in command of concepts today that Aristotle could not even express in his language, Classical Greek. Yes, we have concepts today that would flummox Aristotle – the ideas from atomic and subatomic physics, relativity, the Big Bang, evolution, and so forth. But these concepts would flummox him precisely because he could come to understand them, and they would clash with his view of the natural world. It just doesn't seem likely that we *couldn't* communicate to Aristotle the concept of an atom or evolution in his native tongue (further evidence, included on purpose: the word 'atom' derives from the ancient Greek word for *indivisible*). Though Aristotle did not contemplate atoms or evolution or the Big Bang or the computer, it certainly seems plausible that he could have, using his own language.[2] Word meaning, whatever it is, is therefore an analogue of unchanging reality, while sentence meaning is an analogue of changing appearances. In this way, one of the deepest problems in philosophy, the existence of a grounding reality, emerges as another deep problem, the problem of meaning. The meanings of words somehow interact via context and combine to produce an infinity of new sentences expressing new thoughts.

Before we can relate the problem of meaning to (un)intelligent machines, we need to understand that all of our machines are in fact unintelligent.

3. The Unintelligent Machine

The ultimate goal of artificial intelligence is to build a machine as smart as a human – as smart as you, for example. Many human people, philosophers and nonphilosophers alike, think that this goal is unattainable. The reasons given are often quite basic. One such reason, and a prominent one, comes from a simple observation: humans seem a lot smarter than computers. Not just smarter as in Einstein was smarter than I, or I am smarter than a chimpanzee, but more like I am smarter than a pencil sharpener. Yes, computers can outplay humans at such complicated games such as Go and Chess, and they even play video games. Furthermore, computers can acquire their expertise by 'learning' it while playing. But so what? In 2017, the program AlphaGo was the world's champion Go player – a remarkable achievement perhaps, but far less remarkable than the ordinary daily behaviour of a merely competent human Go player. When the games are over, the human goes off to change a diaper, plan an invasion, drive a car, design a building, write a sonnet, balance accounts, build a shelter, set a bone, comfort the dying, take orders, give orders, cooperate, act alone, solve equations, analyse a new problem, pitch manure, program a computer, cook a tasty meal, fight efficiently, die gallantly.[3] AlphaGo can do none of these, let alone all or even most of them. No machine can. (For more, see Third War Supplement 5.1: Examples of Unintelligent AI.)

Communicating using language is central to being human. Computers are not just notoriously bad at using language, they are laughably bad: just try relying only on voice-to-text on your cell phone (recall the Loebner-prize chatbots of Chapter 2; IBM's Watson, the *Jeopardy!*-winning computer is no exception here – it cannot converse with humans). The fact that computers are bad at using language to communicate with us is a sure sign that they are not intelligent. Dogs in good families routinely know anywhere from 100 to 200 words, sometimes many more. Following the meanings of these words, the body language of their humans, and using their own vocalizations, body language, and facial expressions, dogs communicate better with us than any computer . . . better than any computer on the horizon.

This leaves us with the view that AI is primarily about building useful computational tools to help us humans live better, safer, perhaps more fulfilling lives. Such a view is an accurate portrayal of AI today. However, some hold that the days of AI-as-mere-tool are rapidly coming to an end: computer superintelligence might be just around the corner. And this causes headlines and fear (the former because of the latter).

There is a clash between those who fear AI because of its potential to produce a superhuman intelligence that we could not control and those who think that AI programs are not only not intelligent, but not even on the scale of gradations of intelligence (like pencil sharpeners). Worms are on this scale, possibly even trees, but not computers. Stephen Hawking and Bill Gates are prominent recent AI worriers. In a 2014 interview, Hawking told the BBC, 'The development of full artificial intelligence could spell the end of the human race.' And in 2015, Gates is reported as saying that he was concerned about AI superintelligence.[4] It must be, therefore, that the worriers like Hawking and Gates also worry that the gap between our current unintelligent AI programs and superintelligent ones can be closed, and closed quickly. Such a quick closing of the gap is called *the singularity*. (For more on the singularity, see Kurzweil, 2005, Chalmers, 2010, and Bostrom, 2014[5]). The worriers stoke public fears and get a lot of press, as do programs like AlphaGo, but right now, computers are stunningly stupid, and it is quite unclear how to make them smart. This fact gets little press.

To sum up, the claim in this section is not that computers in the first part of twenty-first century are unsophisticated. They are clearly sophisticated. The claim is that no computer, no matter how good it is at Go or driving, or even driving while playing Go, is as intelligent as a human or even a lobster. Humans are currently the smartest thinkers on Earth. And there are no electronic contenders anywhere on the horizon. (Again, for more, see Third War Supplement 5.1: Examples of Unintelligent AI.)

Could the unintelligence of computers be due to their lack of meaning? The next step in answering this question is showing that our machines are in fact devoid of meaning.

4. The Meaningless Machine

To some humans, the AI deniers, it is obvious that machines cannot think about anything. It's important to be precise here. These particular deniers do not focus on whether computers can have thoughts, hopes, beliefs, dreams, fears, and desires all directed at the relevant things. Rather their denial focuses on whether computations can be *about* anything. Humans of course *do* think about things, all the time and about all kinds of things – from apples and kumquats to prime numbers and transfinite ordinals, to unicorns and ghosts, death and life, and even such things as four-sided triangles. But many hold that mentally grasping such things is impossible for computers.

Semantics, or being meaningful, is the interesting property certain things have of being *about* other things. Words are a paradigm case. 'Dog' (the English noun) refers to dogs (the animals). The notion of being *about* other things is carried over to thinking in a very natural way: mental semantics, *meaning*, *is* thinking about things. The claim then is that because machines (computers) cannot execute computations about things, they are meaningless – they are without any meaningful 'thoughts' – their computations are meaningless to them (but not to us, of course). Just as the words and sentences on this page mean something to the reader, but not to the book the words and sentences are part of, so the computations in a computer are meaningless to it. Computers have no semantics. (Though if one is feeling generous, one can say that computers have *derivative semantics*: their semantics depends on our interpretations of their computations. See e.g. Sloman, 1994.)

The technical philosophy word here is 'intentionality'. Intentionality is the power of a process to be *directed at* or *about* certain things like objects, properties, and states of affairs. The idea is that intentionality is *very special*: only thinking exhibits intentionality; all other physical processes like fire or flowing water or wind are devoid of intentionality. Michael Dyer has a good definition: 'What is intentionality? It is whatever humans have that make them "know what they are doing"' (1990). Most of the war over the semantics of AI machines, or their lack of said semantics, has been couched in terms of intentionality.

The notion of intentionality has a long history. It was probably first worked on well before Plato. A notable example is the philosopher Parmenides, who held that it is impossible to think about what is not. (See the remaining fragments of his book, now referred to as *On Nature*.) Research on intentionality continued up through Avicenna (*c.* 980–1037). Medieval philosophers, such as Aquinas (*c.* 1225–74), John Duns Scotus (*c.* 1266–1308), and William of Ockham (*c.* 1285–1347) vigorously investigated the notion. Then the notion fell off philosophers' radars. Brentano (1838–1917) rehabilitated the term and reintroduced it in his 1874 book *Psychology from an Empirical Standpoint*. Brentano argued that intentionality was *the mark of the mental*: a being can think/has a mind if and only if it has intentionality.

We won't be using the term 'intentionality' here. We will be using 'aboutness' and its cognates. The reason is that 'intentionality' can easily get confused with the word 'intention' (as in doing something on purpose) and 'intension', a term from logic and set theory.[6] So for us, in our terms, a being can think/has a mind if and only if it can think *about* things, i.e. if and only if it has a mental semantics. Since computers' computations are not about things, they cannot think.

Perhaps this major lack in computers isn't obvious to the reader. One quick argument that computers cannot think about things is to consider these three cases: a calculator, a mousetrap, and a thermostat. And then to note that a computer is just like these three in the relevant way. First, the calculator. Type in 2 + 3 and it returns 5. Is it plausible that the calculator is thinking about the numbers 2, 3, 5, and the addition operation and the nature of being equal? It seems not (cf. Dretske, 1985). Yet the calculator unerringly produces the correct output for any input arithmetical operation. How? By following an *algorithm*. Mousetraps don't think of mice. The traps merely operate as they are designed to do when a triggering event occurs (usually a nibbling mouse – but a stick pushing the trigger will work, too). Mousetraps are also following an algorithm. Thermostats don't think about the temperature in one's house. They merely measure the temperature and act accordingly because of the way they are designed. When set to cool, if the temperature gets too high, an electrical circuit in the thermostat is completed which turns on the cooler; when set to heat, if the temperature gets too low, the same thing happens which turns on the furnace. Thermostats, too, are following an algorithm.[7]

Computers of course, also follow algorithms. The algorithms followed are many times more complicated than those followed by calculators, mousetraps, and thermostats. But that is not important. What is important is all four do their jobs by following algorithms. Computers, just like thermostats, etc., don't have to know what they are doing, they don't have to know what their computations are about – indeed, their computations don't have to be about anything for them. Computers simply follow algorithms. Calculators, as we noted above, know nothing of numbers, yet they are very useful for calculating sums, products, and so forth. So, for any device, following an algorithm means that that device doesn't have to know what its processes are about, and indeed its processes don't have to be about anything (except to us).

So, whether the computer is modelling global warming or learning your purchasing preferences from what you search for and buy, it is *not* thinking about you, your preferences, what you search for or buy, global warming, or anything else. Is it plausible that the word-processing program I'm using to write this chapter is thinking about what I'm writing? Is it inwardly objecting to what I am saying about its inability to think about things? Is it thinking: 'Hey, I'm in here thinking about you and how wrong you are!' No, it is not plausible at all. (Apologies to my computer if this is false.)

Let's call this problem, *the problem of meaningless algorithmic processing* ('meaningless processing', for short). Computers apparently suffer from it;

humans and all the other animals on the planet apparently do not.[8] The questions now are 'Why do machines suffer from it, if they do?', 'Why don't humans suffer from it, if they do not?', and 'Can we alter computers so that they don't suffer from it?'

Here is a good spot to introduce one of the most famous anti-AI arguments in the entire AI wars: *The Chinese Room Argument* (Searle 1980, 1990). This argument makes a seemingly strong case that computers completely lack any meaningful symbols.

The Chinese Room Argument

Imagine you are locked in a room. The room also contains a large book plus all the necessities you'll need to be healthy and moderately happy. Strings of marks and squiggles of some sort are written on pieces of paper and slid through a slot in the door. Your job is to look up the strings and individual squiggles in your book, write down on paper other similar strings per the instructions in the book (which, we can suppose, is written in English), and then slide your pieces of paper back out of the slot.[9]

Unbeknownst to you, the characters are Chinese (there is no one language called Chinese; this fact will be ignored). We assume that you know no Chinese at all. (If you do, substitute some other language of which you are completely ignorant with an alphabet different from any language you know.) You go about your job, laboriously reading and following the instructions in the book, writing the characters, to the best of your ability, on sheets of paper, and sticking the sheets out through the slot. This goes on for months, during which time you are otherwise well cared for.

Also unbeknownst to you, outside the room are several Chinese scholars discussing, say, Chinese history in Chinese. Because of the insightfulness and erudition of the comments coming out through the slot, they believe that the person in the room is an expert on China's history as well as being fluent in Chinese. You are not an expert on Chinese history; worse, you are completely ignorant of what language you are working on. Indeed, you entertain the idea that the symbols you are manipulating are not in a language at all. The only reason you think they might be is that the book of instructions seems so complete and organized, even if ponderous. So, outside the room, an erudite discussion of Chinese history, inside the room, a tedious manipulation of meaningless symbols.

Here is where Searle springs his trap: you in the room with your book and sheets of paper are *computing*. In all relevant respects, you are implementing a computer program – a program for communicating in Chinese about Chinese history. In fact, the large book in your room contains the algorithm (or implements it in English). You are simply following the algorithm (again, unbeknownst to you). Since you are computing (i.e. following an algorithm) and since you have no idea what you are 'talking' about (Chinese history), it follows that neither would a real electronic, digital computer, implementing the same algorithm you are implementing by using that large book.

Since there is nothing special about Chinese history, the argument metastasizes. The essence of computing, so Searle's argument says, is doing exactly what you are doing: algorithmically manipulating mere marks without regard to their meaning. Hence, computing is *not* processing contentful symbols. Computing is contentless manipulation of marks (in a computer, these marks are electrical signals). But we noted above that thinking is processing content – all your thoughts have content, and the content they have is crucial to the way they are manipulated and processed. Ergo, computers can't think. Since thinking is a necessary condition for intelligence, computers can't be intelligent: the AI project is hopeless.

The Chinese Room Argument engendered multiple replies and objections, several summarized in Searle's original 1980 paper or published with it, together with further discussion by Searle, in the journal *Behavioral and Brain Sciences*. Many papers arguing one side or other have been published since. We review the 'standard' replies in Third War Supplement 5.2.

Intuition and the Chinese Room Argument

It is important to note that both arguments above, the calculator-mousetrap-thermostat argument and the Chinese Room Argument, rely crucially on *intuition*. We have a clear and distinct intuition that in following an algorithm, e.g. thermostats are not thinking about the temperature. Without this intuition, the argument cannot be made. But intuitions are *perceptions* (see Chudnoff, 2013), and are, therefore, (1) immediate, or at least not mediated by rational argument or conscious reasoning, and because of (1), (2) intuitions are not derived from a body of supporting data which establishes their truth; they are *perceived truths*. In believing that thermostats are not conscious or accepting Searle's argument, we are taking our intuitions seriously as sources of knowledge. (Some of the trouble using our intuitions often gets us into will be discussed in Chapters 6 and 7.)

Finally, we stress that the information is meaningless, in these arguments, only to the processor processing the information: the calculator, the mousetrap, the thermostat, the computer, you-in-the-room. The information is definitely *not* meaningless to us humans – this is why we built the devices and machines in the first place. So, computers work without their computations being about anything to *them*, but they are about something, derivatively, to *us*. All the more so, computers neither know nor care what they are working on, what their computations are about for us. In fact, it misses the mark to even say that computers don't know or care what their computations are about. It is a little like saying that a rock doesn't know or care that it is heating up in the noonday sun: the rock isn't the sort of thing that *can* know or care; it just heats up. Same with computers: they just execute algorithms.

Interlude: Dretske on Meaningless Processing vs. Rapaport's Meaningful Processing – Syntactic Semantics

This a good place to review an important battle in the AI War over meaningful/less computational content. The philosopher, Fred Dretske, in his 1985 paper, made a particularly strong attack on AI by attacking the aboutnessless of computers. He said that computers, including calculators, cannot even add. Why? Because having internal representations, data structures, is not sufficient for aboutness. Dretske said: '[Computers] don't solve problems, play games, prove theorems, recognize patterns, let alone think, see, and remember. They don't even add and subtract' (1985, p. 24).

One has to give credit to Dretske for following through: If computers cannot think about things, then, as noted here, they can't add, since adding requires thinking about things – numbers, to be exact. Most other philosophers had failed to follow the 'computers can't think about things' objection to its full conclusion – most philosophers did think computers could add. But if Dretske was right, then how do humans think about numbers? This is an issue to which he is keenly sensitive (p. 26ff). Dretske's solution to this problem is the *robot reply* to Searle's Chinese Room Argument: A robust mobile, learning robot equipped with sensors and effectors (arms, hands, legs, or some other way of moving around) will, by the very fact of being such a robot, have meaningful symbols (for the Robot and other replies to the Chinese Room, see Supplement 5.2 to this chapter). So, a learning robot can add, a human can add, but a computer cannot. (See section 6, below, where we will examine a robust mobile learning robot to see whether it has any aboutness).

In what is arguably the strongest push back by an AI researcher against both Searle and Dretske (among others), Rapaport argued via his theory of *syntactic semantics* that computers can rather easily have genuine aboutness. Rapaport develops and presents his theory in many publications, the central ones being his 1988, 1995, 1999, 2000, and 2006. Rapaport first pushed back against Searle. In his famous paper, Searle says:

> Because the formal symbol manipulations [computations] by themselves don't have any intentionality [aboutness]; they are quite meaningless; they aren't even *symbol* manipulations, since the symbols don't symbolize anything. In the linguistic jargon, *they have only a syntax but no semantics.*
>
> emphases added; and see below, section 5

What Searle means is that computers can only use rule-governed transformations to change one data structure into another, all without knowing what the data structures stand for or represent. Rapaport argues that such rule-governed transformations *are* semantics, a *syntactic semantics*, or a semantics emerging from syntax. In short, Rapaport argued that a machine could have aboutness (semantics) merely in virtue of syntactically manipulating data structures. This push back will also work against Dretske, Rapaport argued. Here is how syntactic semantics works.

Let's assume that there is a computer, called AlphaZero (see the Third War, Supplement 5.1, below), which knows a lot about chess and can play chess so well that no human no matter how good can beat it. When AlphaZero is playing, there are three things, domains, that we are interested in (c.f. Rapaport, 2006):

1 AlphaZero's internal computations and data structures, which mean something to us, but not to AlphaZero. It is assumed, required in fact, that most of these internal computations, though of course syntactic (they are the result of computations, after all) are semantic on their own. Why grant this? Because the internal computations are mostly made up of symbols that stand for other symbols inside of AlphaZero. In particular, many parts of AlphaZero's running program refer to other parts of that same program. The simplest case of this is variable binding: If X is bound to 1, then when X gets incremented by adding 1 to it – X becomes X + 1 – AlphaZero has to be able to retrieve what X is bound to in order to perform the computing that involves X. As a result of the binding and the increment, X is now bound to 2. In this way, AlphaZero has access to the interpretations of most of its internal symbols. (We say 'most of' because AlphaZero does *not* have access to

the interpretations of its symbols for 'chess board' and 'chess pieces'. It is for this very reason that AlphaZero is said to lack aboutness.)

2   The external chess board, out in the world, on which stand the chess pieces in various configurations specified by relations to one another.

3   A set of correlations *we use* from the internal computations to the chess board and pieces. This correlation is part of what allows AlphaZero to win at chess via making internal computations.

The key to Rapaport's idea is that the external chess board and all its pieces can be *internalized* – implemented inside of AlphaZero. Now, AlphaZero can play chess without a physical chess board. Everything important that is occurring is occurring inside AlphaZero. Rapaport calls this *Internalized Meaning*, and says that such internalization is 'pushing the world into the mind' (Rapaport, 2006). It is important to see that the correlations, i.e. (3) above, are also internalized once the chess board and its pieces are pushed into AlphaZero. It is these correlations that constitute the ultimate semantics of AlphaZero's computations. Before the internalization, AlphaZero did in fact lack aboutness for those symbols that referred outside of it to the chess board and pieces. But now with everything inside, all the semantics is available to AlphaZero (basically, the correlations in (3) supply AlphaZero with the extra variable bindings, as discussed in (1)). Of course, this internal semantics is still syntactic because it is now all one large computer program. Rapaport finishes his theory by arguing that we natural thinkers have aboutness simply because our sense organs internalize the world for us. So, in an important sense, Rapaport's theory of machine aboutness is an unusual version of the Robot Reply to Searle's Chinese Room Argument. Aboutness comes into existence when all the semantics is syntactically internalized. We are living robots who have aboutness because our environments are appropriately internalized.[10]

End of Interlude.

5. Must the Meaningless Machine be Unintelligent?

There would be little use in railing against AI because computers are meaningless if there were no connection between meaninglessness and being unintelligent. Oddly, the connection between the two has rarely been explicitly spelled out. Why is this? The short answer is that either the connection was thought to be obvious, or the connection was completely obscure, so all naysayers ignored it.

At the end of the day, there are only two viable approaches to aboutness. On one view, aboutness is just being able to manipulate objects and other parts of our environment – food, for example. On this view, aboutness in machines will be achieved with the development of robust *robots*. And obviously, on this view, aboutness is necessary to being intelligent. However, on another, possibly more appealing view, aboutness is *consciousness*. Of the many deep difficulties surrounding consciousness, one is that it is not known whether or not consciousness is important to being intelligent, and if it is, it is not known why. (It should be noted that the robot approach assumes that consciousness is not needed for intelligence, and that consciousness is at most a side effect of being able to get along in a natural environment. We return to this in section 10 and the sections following it.)

Searle (1980) argued that being meaningless meant lacking the proper causal connection to the world. So, Searle solved his Chinese Room problem by invoking causal connections. Real thinking things – humans, cats, dogs, etc. – have this connection. Searle says:

> 'Could a machine think?' My own view is that only a machine could think, and indeed only very special kinds of machines, namely brains and machines that had the same causal powers as brains.
>
> 1980, p. 424

So, we and the other animals are smart because we have the right causal powers. This seems to be an intuitively appealing explanation, at first: computers lack a causal connection to the world humans and other genuine thinkers have. According to Searle, this lack serves to *both* make computers unintelligent *and* to render them meaningless. The real key, then, is lacking the right causal powers (see Supplement 5.3 for difficulties with this causal picture).

Unfortunately, Searle never spelled out what 'causal powers' brains have that set them apart from computers. And this missing definition has never been supplied, not by Searle, nor anyone else.

To summarize: Humans and other animal minds appear to have two crucial properties: intelligence, and aboutness. We have a crude definition of the first one ('able to solve problems as well as a human' will do). But we have no agreed-upon definition at all of aboutness. So the AI War about mental semantics was fought without anyone really knowing what they were fighting over. The truth is, there is no consensus on why computers and robots are not very smart. Lacking an argument explicitly linking aboutness to intelligence, it is just a guess that the former is crucial to the latter.[11]

To see better the problems of aboutness and intelligence, let's now examine the two main ways to solve the aboutness problem. The first way is widely regarded among AI researchers and friends as the best way to solve this problem. This way puts all its money on *robots* – on being able to move around and having sensory mechanisms (cameras and hearing devices, etc.) and effectors (things like arms and fingers and the like). The second way is widely regarded as hopeless. This way involves *consciousness*. The perceived hopeless of this view is hard to overstate. Many researchers have the attitude that the less said about consciousness the better. The well-known philosopher and cognitive scientist, Jerry Fodor, once said, 'I try never to think about consciousness. Or even to write about it' (1998, p. 73).

6. Aboutness as Interacting with the World: The Robot Pencil Sharpener

The point of this section is to examine the claim that aboutness is *fully* captured by interacting with the world: Only robots can be meaningful machines.

Assume that you have a standard pencil sharpener, and that your pencil sharpener cannot think about pencils (as it surely can't). It is merely a tool for sharpening pencils. Now suppose you set out to make an intelligent pencil sharpener. What would you have to do?

One problem right off is that a standard pencil sharpener has no means for thinking about pencils. To think about pencils, the sharpener would need some sort of *physical internal representation* that designates pencils for it, and (this is crucial) this internal representation would have to be *manipulable* by the sharpener. 'Manipulable' means two separate things. First, it means that the internal representation would have to be involved in well-known mental processes like inferences ('if the pencil is sharp, there is no point in sharpening it'), generalizations ('All pencils need sharpening eventually'), and categorizations ('That's not a pencil'). (This list is only partial. These mental processes, especially inference, are collectively known as *reasoning*.)

However, note that all these processes involve *other* internal representations that the pencil sharpener would have to know: that *sharpening causes sharpness*, this requires in turn the notion of *causation*, and then the notion of *eventually*, the notion of *being dull*, the notion of *not being a pencil* (and so the notion of *being a pencil*), and so on and so forth. (In cognitive psychology, all these internal representations are called *concepts*.) The requirement of needing to be manipulable is why the spiral metal sharpeners in your standard pencil sharpener *cannot* be used as the internal representation of pencils: the metal sharpeners are not mentally manipulable, so they do *not* allow your pencil sharpener to think about pencils; they merely allow the sharpener to sharpen pencils.

Second, being manipulable also means taking part in different kinds of thoughts.[12] Suppose you get your sharpener to *believe* that all dull pencils need sharpening. This is the sharpener's reason for being, so this belief must be included, and it has to be foundational. Let us grant that representations involved in inferences, generalizations, and categorizations are believed. But believing is *not* sufficient to get your robot sharpener to sharpen pencils. The only way to do that is to get it to *want to* or *desire to* sharpen pencils. So, your robot has to have beliefs and desires (wants). But it also has to have more that these two kinds of thoughts. It also needs to be able to *intend*, to *plan*, to *doubt*, to *disbelieve*, to *suspect*, to *suppose*, and on and on.[13]

With just this much, we now have three conclusions:

1 Your intelligent pencil sharpener will have to be able to think about pencils, but
2 To do (1), your sharpener will have to think about a lot more than pencils. And,
3 To do (1) as well as (2), your sharpener will have to do many different kinds of thinking, including, but not limited to believing, desiring, hoping, intending, planning, doubting, and suspecting.

We see, then, that thinking about pencils and usefully sharpening them requires a large collection of different kinds of thoughts and thought-contents. But this is only the beginning.

Let's assume that you can somehow satisfy (1). How are you going to satisfy (2)? By using the same method you used to satisfy (1), plus something extra ... The list of other concepts that you will have to 'give' your pencil sharpener is not only large, but open-ended, meaning you don't know exactly what extra concepts are needed, or how many are needed. This problem of open-ended additions is standardly solved in AI by *learning*. So, you'll need to get your pencil sharpener to learn (there are many learning algorithms available). It will have to be able to add to its store of concepts on its own without your help.

How are you going to satisfy (3)? It is fair to say that no one knows the answer to this question. It does appear as if most, if not all, kinds of thinking are innate – that is, infant brains are hardwired with them. However, there could well be some sort of developmental process that increasingly separates believing from desiring, intending from planning, etc. But again, no one knows what this process is. So, your project seems dead right here. Undaunted, you press on. You assume what is called a *functionalist approach* to the different kinds of thinking: what makes a thought the kind of thought that it is is the *role* it plays in the overall cognitive life of the thinker.[14] A thought is a belief if it functions in inferences and as a guide to knowledge and action, a thought is a desire if it functions to cause or direct actions, a thought is a doubt if it functions to dampen other beliefs and desires, and so on and so forth. The great thing about the functional approach is that it makes all thought processes programmable. Armed with this, you set off to implement a computer capable of different thoughts. Your pencil sharpener is a complex programmable, learning computer that sharpens pencils. Yes, there are easier, cheaper ways to sharpen your collection of pencils, but you are burdened with glorious purpose way beyond mere writing utensils.

But what of (1)? How do you get your fancy sharpener to think about pencils (or anything else) in the first place? One well-researched way is to add cameras as well as arms and grabbers of some sort to your pencil sharpener (again, there are many kinds of all three available). Adding programmable cameras allows your sharpener to find the pencils it needs to sharpen. Adding arms and grabbers (hands), allows your machine to grasp the pencils it has located with its 'eyes'. The arms and grabbers will also need sensors to detect if a pencil has been successfully picked up, etc. Now you have something resembling a *robot pencil sharpener*. It can use its cameras and categorization ability to find dull pencils within its arm's reach, pick them up with its grabbers, and sharpen them. If you add legs or wheels for mobility, then the limitation of having to be in arm's reach can be avoided. Of course, you'll also have to add sensors so your robot can tell where its legs and arms are and where its cameras are pointed. And lastly, you'll need to implement some internal way for your robot to know when a pencil has actually been sufficiently sharpened and to then stop sharpening it.

But after years of work, you have all the necessary ingredients implemented. Now your robot pencil sharpener is ready. For a test, you begin writing with a pencil. When it gets dull, you toss it aside, you know not where, grabbing another pencil. Your clever robot pencil sharpener scurries over to where the pencil fell, picks it up, sharpens it to the required sharpness and brings it back and then lays it gently on your desk. And, let us suppose, it does all this without getting in your way or tripping over books and papers you have lying around. You are now never without a sharp pencil. You become moderately famous and rich.

7. Some Lessons from the Robot Sharpener

We see, therefore, that the task of getting your sharpener to think about pencils expands, and must expand, into getting it to learn and use many other concepts, getting it to have many kinds of thoughts, and into making a robot, not just a stationary computer with a pencil sharpener. There is, as it were, a sort of minimal mental (cognitive) 'size' your robot has to be to actually be intelligent, even in the narrow realm of sharpening pencils. It will have to learn, not only about pencils, but how to use its legs and arms and sharpener effectively. And it will have to think about learning and moving and grabbing . . . And on and on. (Matters only get more complicated if at some point, you decide you want to implement a way to converse with your robot sharpener.)

(We also stress that we have purposely avoided discussing adding *emotions* to your robot. Quite plausibly your robot will have to have emotions. Why? Merely desiring to sharpen pencils is unlikely to be enough. Your robot's desiring will have to have some sort of emotional component to push it to sharpen pencils. Simply put, reasons alone cannot power the will. The Scottish philosopher David Hume compellingly argued for this idea. See Book 3 of his 1739–40. It is impossible to overstate how much emotions complicate matters here. This puts your project of building a robot pencil sharpener relying only on reasoning, learning, and moving around in the realm of science fiction, since it is extremely unlikely that Hume was wrong. We will take up one version of this issue again in 'Mattering Requires Consciousness', in section 13.)

The result is that your robot pencil sharpener must acquire a large *web* of interconnected concepts, beliefs, and knowledge. This is interesting. By following the AI story about your robot sharpener, we can see that thinking about one thing is not possible: to think about one thing requires thinking about many. But we also know that thinking about everything is not possible (at least it is not physically possible). Hence, your sharpener (which is physical, of course) must think about a *collection* of things that constitutes a small subset of all the things there are in our universe. And your robot must be able to add to this collection.

So, to be a thinking thing, your robot must inhabit a *cognitive environment*, a *cognitive world*. Such a world is the collection of things from the external world that your robot can and does think about and physically manipulates.

Finally, it is crucial to note that this cognitive world is expandable. Once it is up and running, your robot pencil sharpener might be able to do other things for you beyond sharpening pencils. It might be able to retrieve pens, books, lunch . . . the sky's the limit.

Conclusion: Your robot pencil sharpener certainly *seems* to think about pencils. It can, autonomously, find them, sharpen them, and put them where they are accessible to you as you work, all the while avoiding obstacles. The robot sharpener seems to exist in a complex, pencil-sharpening world. Let's explore this conclusion.

8. The Physical Symbol System Hypothesis

In the above section, we have replicated, and given an argument in support of, an important and foundational hypothesis in AI. This hypothesis relates meaning and computing. AI starts off by assuming that cognition, thinking, is a form of *information processing*. AI then moves up from there. This is codified in the *Physical Symbol System Hypothesis* briefly introduced in section 2.3 of Chapter 4:

> The Physical Symbol System Hypothesis
> *A physical symbol system has the necessary and sufficient means for general intelligent action.*
>
> Newell and Simon, 1976

This hypothesis turns out to be the dream of AI in fourteen words because Newell and Simon define a physical symbol system to be a computer (see their 1976). In 1976, this was both a bold and a plausible hypothesis. But it is neither today. Moving beyond Newell and Simon's definition, we can say that a physical symbol system is any physical system that inhabits a robust *cognitive world* by physically using realized symbols of that world (internal, manipulable representations of it) to autonomously accomplish real, physical tasks relevant in that world, like, say, finding and sharpening pencils (or food and mates, speaking more biologically). So a physical symbol system, as defined here, is much more than a mere computer; it is an autonomous agent.[15] Using our definition of physical symbol systems, not all computers are physical symbol systems, but perhaps all physical symbol systems are computers (many researchers dispute this last claim, see the Second War).

9. Aboutness is Doing

It is important to see that concluding that your robot pencil sharpener thinks about pencils rests on the fact that the robot *does complicated things with and for pencils*. We call this *aboutness is doing*. This is a classic response to the problem of meaningless algorithmic processing.

AI researchers have responded to the problem of meaningless processing with three main proposed solutions:

1   Aboutness is doing. This solution is embraced by those who opt for the Robot Reply to the Chinese Room Argument (see Supplement 5.2). Those who side with what is called *embodied cognition* also embrace, or can embrace, this solution (see the Second War). Embodied cognition expands on the idea that aboutness is doing. The embodiment hypothesis is that *thinking* is doing, and this includes aboutness. Sometimes the embodied cognition hypothesis is supercharged by coupling it with the idea that thinking also requires a *non-computational dynamic system* (again, see the Second War). This claim is further expanded to include the idea that brains are such non-computational dynamic systems. The best place to learn about this topic is Spivey, 2007; for a more radical version, see Chemero, 2009, and for critiques, see Markman and Dietrich, 2000; Dietrich and Markman, 2001 and 2003.

2   AI as an engineering discipline is generally unconcerned with whether or not computers can really think about things. Whether they can really think about things is a mere philosophical conundrum of no practical importance. The large field of applied AI lives here.

3   That humans think about things has been completely overblown. All there is to thinking about things is executing an algorithm in which the relevant things are *represented* internally in computational data

structures. So, yes, calculators do think about numbers – and thermostats think about temperature, and mousetraps about mice, for that matter, or at least mousetraps think about appropriate triggers. This is unintuitive, but science is often unintuitive. (This is the view of Newell and Simon, 1976, and McCarthy 1979 – though they differ on why. Newell and Simon have a weak notion of a physical symbol system, and McCarthy relies heavily on pragmatics. Those who favour a version of the Systems Reply to the Chinese Room Argument also go here. Again, see Supplement 5.2)

Solution (1) is a substantive and important thesis. It gives us the conclusion that being a robot is *sufficient* for aboutness. This is obviously quite strong. But it is a clean and readily available solution. Plus, robots are needed in our world culture, and since robust robots are proving difficult to implement, solution (1) points to a strong and useful research topic, linking the semantics of thought with being a robot.

Solution (2) is eliminative: it says there was never an important problem with mental semantics, though, yes, whether computers really think about things might be fun to discuss over coffee one afternoon. This solution is certainly painless and expeditious. But it also seems wrong. Clearly humans and all the other animals think about things. And thinking about things certainly seems important to interacting successfully with the things we think about – as discussed in section 3, computers are not very smart; being unable to think about things might be the culprit (though, as we discussed in section 5, this is just a guess).

Solution (3) says that we've already solved the problem, which wasn't very deep in the first place. Basically, the problem was solved with the invention of the computer. But this solution raises the question: Do abacuses think about numbers? If not, why not? The only difference between an abacus and a computer is that the computer uses electricity to do some of its operations itself. However, an abacus can use gravity to a small extent to help slide the beads, as well as the force fingers generate to quickly push the beads. So again, why draw a sharp distinction between abacuses and computers? Furthermore, do books think about words? Or do they think about what their sentences mean? Answering *Yes* to all these question seems facile. Answering *No* to all these question requires a real theory of aboutness, and thereby returns us to solution (1).

One can ask, however, if aboutness is completely captured by doing. Clearly our (imagined) robot pencil sharpener seems to be thinking about pencils, finding the dull ones, sharpening them, and so forth. It certainly does look like *doing* is all there is to *aboutness*. But is seeming to think really thinking? If something is *ascribed* thought, does it really have thought? For McCarthy (1979) the answer is yes – this is the essence of his pragmatic view. But *cartoon characters* seem to think. Surely McCarthy wouldn't ascribe thoughts to them! If we accept that aboutness is doing, then we won't be able to draw the distinction between seeming to think and really thinking, between seeming to think *about* things and really thinking *about* things. If we adopt 'aboutness is doing', we will be forced into pragmatics: the only aboutness is *ascribed* aboutness, and aboutness is ascribed when it is *useful to do so*. However, if we focus on ourselves for a moment, aboutness definitely seems to be more than doing – much more. Your thoughts about this book you are currently reading are about *this book which are you consciously seeing and reading* and not some ice cream in your fridge or some rocks on the moon. And your thoughts about this book are definitely not meaningless, especially to you. Furthermore, you *know* all of this; you are *certain* of it. This certainty is supplied by your consciousness. In the next section, we begin examining an argument that aboutness is in fact much more than doing – *aboutness requires consciousness*. What we will learn is that there is a non-pragmatic sense of doing that is much more important than AI scientists have thought. This sense requires consciousness essentially. In this sense, your robot pencil sharpener is not doing anything at all, just as Dretske said. We will have a specific suggestion of what meaning, aboutness, consists of: consciousness.[16]

10. Consciousness: What is it?

What has been absent *but oddly not explicitly missed* in all our discussion of aboutness so far is *consciousness*. This is true not only here, but in the history of AI (and all of cognitive science). AI, both during the Wars and now, eschewed consciousness like it was the plague. The conscious machine, the conscious computer, the conscious robot … all these notions were almost forbidden research topics (but see Supplement 5.5 for a review of a recent skirmish). It seems very odd that the study and imitation of the mind would ignore something so obviously important as consciousness. We must now delve into this. (See Chapter 7 for a deeper examination of consciousness than the present section. Here, we only need enough to relate consciousness to meaning.)

What is consciousness? What do we mean by the word 'consciousness'? Here's a 'definition':

Consciousness is the way the world seems to us, the way we experience it, feel it. Taste an onion, see a rainbow, smell a dead skunk on a hot summer's day, stub your toe on the foot of the bed frame at 4:00 A.M., hear your dog breathe or a baby gurgle and coo. These are experiences, bits of our phenomenology, and it is these experiences that somehow give us our subjective point of view . . . We have experiences because we are conscious. Or, rather, our having them constitutes our being conscious. Being conscious is what makes it fun or horrible or merely boring to be a human. Using the phrase that Thomas Nagel (1974) made famous, we can say that a being is conscious if *there is something it is like to be that being.*

Dietrich and Hardcastle, 2004, p. 5, emphasis added

Part of the reason consciousness was (and is) eschewed is revealed in the paragraph following the one above:

[This] ostensive definition of consciousness, appealing to something we can only assume our readers have, will have to do, for there is no more robust scientific, third-person definition available. And this ostensive definition is the same one researchers have been using to define consciousness since research of any sort was first conducted on it.

p. 5

All of science is based on robust *third-person* definitions of the various technical terms: mass, charge, spin, electron, gravity, DNA, protein, cell wall, species, adaptation, acid, base, metal, computation, etc. But in this one case, consciousness, such a third-person robust definition is not possible, or at least, has not been adopted outside of the narrow confines of medical practice (see Chapter 7). Only a first-person definition is available. To get a third-person definition, we'd have to grasp consciousness from a point of view available to everyone. But there is no such point of view. From the 'outside', direct access to another being's consciousness is impossible. However, science is *essentially* public: *science is a kind of social knowledge.* Science, therefore, cannot use a first-person definition. So it is bad news that any definition of consciousness *must* isolate it to the 'inside' of one individual: the conscious being.

There is another deep problem with defining consciousness. We quote Dietrich and Hardcastle (2004, p. 8):

Scientists have hypothesized many intriguing and insightful things as being identical to consciousness. Here are some of them: attention, autobiographical memory, being awake, body-based perspectivalness, neural competition, episodic memory, executive processing, feedback, feature integration, 40 Hz neural oscillations in human brains, high-level encoding, intentionality (as in intending to do something), intentionality (as in a mental representation's being about something in the world [see section 4, above]), meta-processing, mind-based perspectivalness, quantum effects in the microtubules of neurons, recursivity, reflective self-awareness, reportability, salience, sense of self.

That is quite a list. Note how far and wide it ranges. Some of these suggestions reduce consciousness to brain processes (40 Hz oscillations, quantum effects in microtubules), others attempt to identify consciousness with some psychological process or property (attention, executive processing, various memory systems), while still others move the problem of consciousness to some other cohort property of equal rank (intentionality, reflective self-awareness, subjectivity, sense of self). None have succeeded. All the items on the list have one of two properties: they either are necessary (at best), but not sufficient, for consciousness, or are as puzzling as consciousness itself. All of the attempts to reduce consciousness to brain processes or to psychological processes give us necessary properties for consciousness (perhaps). All the attempts to explain consciousness by using some mental property of equal rank merely replace the problem of consciousness with one equally puzzling.

So, we have a very rough 'definition' of consciousness, which is as good as anyone can do. Hopefully, our reader agrees with this definition, and even more hopefully, agrees because she or he is in fact conscious. Now to aboutness and consciousness.

11. The Relation Between Aboutness and Consciousness

Your thoughts, of whatever kind, are about things because you are conscious. More specifically, your thoughts are about what you are conscious of thinking because you are having those very thoughts. The same principle governs perceptions: your conscious perceptions are about what you are consciously perceiving and attending to. Some examples will help explain all this.

Suppose you are looking at and attending to a pencil, and that you are thereby conscious of the pencil. Your perception is *about* that pencil. If you think 'That pencil is sharp' then your thought (which is a perceptual belief) is about that pencil and its sharpness. The philosopher David Chalmers calls this kind of thought a *first-order* judgment. If you move up a level to a *second-order* judgment and thereby attend to your perceiving the pencil, then you are conscious of, not the pencil only, but of perceiving the pencil.

So, in this case, your first-order perceiving is what your second-order thought is about.[17] If later, as you hike along some trail, you remember the pencil, then you are conscious of the pencil again, only this time, through memory; you are conscious of the pencil as perceived through your *mind's eye*. Again, your thought is about the pencil, even though the pencil is absent.

On this view, something that supposedly thinks but is not conscious never has thoughts about things (if it even has thoughts at all). Such a thing merely has useful behaviour. This is true whether such a non-conscious thing is a robot or not. If the robot sharpener isn't conscious, then its perceptions and computations are meaningless (but useful). And if you have an unconscious thought, it is not about anything either. However, a thought (or a computation) that is not about anything is not useless. It could well be something that crucially supports an occurrent conscious thought or one that will be conscious in the future.

So, your thoughts are about what you are conscious of them being about. If we define *semantic content* to be what a thought of yours is *about*, and *conscious content* to be what you experience by having a certain thought, then, on the view argued for here, we get the following equation:

Semantic Content = Conscious Content.

Very tidy, and presumably obvious to anyone or anything that is conscious and doesn't have a theoretical axe to grind. But this is not an equation embraced by the artificial intelligence and cognitive science communities . . . to put it mildly.[18]

12. A New Problem: What is Consciousness for? The Standoff

However now we have a new problem: what work is consciousness actually doing? What, specifically, is the difference between your aboutness-is-doing robot pencil sharpener and a new robot sharpener that is conscious of what it is doing? (Remember, we are assuming that the aboutness-is-doing robot sharpener is *not* conscious.) This question expands: In humans and other animals, what is consciousness for, and how does it contribute to our being intelligent? One would think the answer to this question is obvious, but it is not. The problem is the possibility of *zombies* – see the Third War Supplement 5.4: Problems with Consciousness.

The argument starting from section 6 has produced this tentative conclusion:

(Tentative conclusion) The robot pencil sharpener proves that aboutness is doing. In proving this, it also proves that nothing else is needed – specifically, being conscious is *not* needed for intelligence.

There is a problem with this conclusion, however.

It is a *gift* to the robot pencil sharpener thought experiment to accept that the robot sharpener would work well even though it lacks consciousness – that it doesn't need to be conscious to work. (It is also a huge assumption that it is *not* conscious – an assumption we will continue to make for now. For more on this assumption, again, see the Third War Supplement 5.4: Problems with Consciousness.)

Technically, assuming that the robot pencil sharpener would work even though it is not conscious is a form of the fallacy of *begging the question* (assuming what you want to prove). Why? Because were you to actually see a robot pencil sharpener working in someone's study or lab scurrying about, grabbing and sharpening dull pencils, returning them to the right person and place, dodging people, books, and papers on the floor, navigating around desks and ordinary computers, you would find it impossible to believe that the robot wasn't conscious. Indeed, it would be natural to think that the robot was conscious. To deny this on the basis of merely imagining a non-conscious robot sharpener is to assume that consciousness is not needed for intelligence, and this is exactly what is at issue.

So, the robot sharpener and hence the tentative conclusion depend on committing this fallacy. This is enough to make us doubt, at least to some degree, the tentative conclusion.

Well, if conscious is needed, what does being conscious allow the robot to do that a non-conscious robot couldn't? Answer: Effectively find and sharpen dull pencils, and then return them; all without being in the way or tripping over something.

The trouble with this glib answer is we can't prove it – ever. As we saw when we defined consciousness in section 10, consciousness is entirely private. All we can ever have is outward behavioural 'evidence' that inwardly some intelligent being is conscious. And this isn't good enough for science, at least usually (perhaps science will have to change when it comes to consciousness; see Chapter 7 for an extended discussion).

Put it this way: Presented with the highly useful robot sharpener, all that we could ever know about it is that it worked — that its behaviour was exactly what pencil users of the world wanted. And we know *why* the robot works because we know how its program works and we know how its various physical parts work. The further deeper claim that the robot *really* works because it is conscious and its thoughts are really about things in its environment is forever unprovable by us or anyone. This deeper claim, therefore, becomes merely an article of faith.

So now we have a standoff. The 'aboutness-is-doing' crowd has no business assuming that the robot sharpener is not conscious – that it can work without being conscious. But the 'aboutness-is-consciousness' crowd has no business assuming that the robot sharpener needs consciousness to work, nor that AI in general needs conscious computers to finally produce a genuine intelligent, thinking machine. (It must be stressed that the 'aboutness-is-consciousness' group was a relatively small but widely varied group of philosophers. This group is usually classified as belonging to the larger group called 'intentionalists', who hold that consciousness and intentionality should not be separated, though some intentionalists did *not* in fact take consciousness seriously. Also, the aboutness-is-consciousness group contained no AI researchers. For much more, see Siewart, 2017.)

13. Resolving the Standoff: Two Jobs for Consciousness

But perhaps the situation is not as dire and shrouded in mystery as it seems. If consciousness had a specific and important job to do in our overall cognitive economy, then we might safely infer (as in *inference to the best explanation*[19]) that anything with a robust cognitive economy must be conscious. Such an inference would not be a fallacy, it would be, as just noted, a justifying explanation.

Are there any candidates for such a job? Yes, there are two: knowledge and mattering. We will explain these in order.

Knowledge Requires Consciousness

What if consciousness is required for us to know things? To develop this, let's first distinguish between *knowing-that* and *knowing-how*. This distinction is common in philosophy, and heavily theorized about. It is old, too, going back at least to the ancient Greeks who distinguished between *epistêmê* (typically translated as 'knowledge') and *technê* (typically translated as 'trade' or 'craft' or 'art'). Knowledge-that is when you know facts like: Alex Honnold was the first person to climb, free-solo, El Capitan in Yosemite National Park. Knowledge-how is skill, like Honnold knowing how to climb or you knowing how to throw a ball or ride a bicycle.

Since Gilbert Ryle's important treatment of the topic (ch. 2 of his 1949), most philosophers believe that knowledge-that is to a large extent independent of knowledge-how. Here, we will focus exclusively on knowledge-that, relying on the plausible independence between the two to guarantee that we are not leaving out anything important (again see Chapter 7, where this assumption is questioned).

We are focusing on knowledge-that because we seem to be mostly conscious of facts, of knowledge-that. Knowledge-how seems to elude consciousness. If you and your friend throw a baseball back and forth, then yes, you know how to throw a baseball, and you are conscious of your throwing. But you are conscious of the *fact* that you are throwing a baseball, you are not conscious of *how* you do it – you know how to throw a baseball, but you do not know how you manage to successfully complete a throw – you just throw. You don't know how you successfully ride a bicycle – you just ride it. Same with speaking and talking with someone. One way to see this is to note that when you teach someone how to throw a baseball, mostly you encourage him or her to throw the ball after watching you throw the ball. You cannot teach them by telling them. You can point out errors in movement, e.g. 'your hand passes too close to your head', you might say, but if it helps, it only helps slightly – the only way to learn to throw is to throw and observe others who already know how. One can learn that Alex Honnold was the first person to free-solo El Capitan in Yosemite National Park by being told. One doesn't have to learn this to climb; one learns to climb by climbing, just like throwing.

Now to hook up consciousness and knowledge of facts.

Consciousness is unique among our mental powers. As Chalmers and others have pointed out, we have unmediated, direct knowledge that we are conscious. Chalmers says: 'Conscious experience lies at the center of our epistemic universe; we have access to it directly' (1996, p. 196). But, necessarily, consciousness consists of contents – all consciousness is consciousness *of* something: a tree, a rock, a cloud, a feeling, an emotion, a vision, a perception, a thought, a hope, a desire, a belief, and so on and so forth. Therefore, we have knowledge of conscious contents directly, unmediated by inferences of any sort. Now, of course, our perceptions might be wrong. But we cannot be wrong that we are having those perceptions. For example, we might see that the full moon is larger when it is rising than when it is overhead (the famous *moon illusion* – the differences in perceived sizes is usually quite large, often the horizon moon looks around 50% bigger than the zenith moon). We might conclude that the moon changes sizes or is closer when it is rising. Investigation reveals that this phenomenon is in fact an illusion: the moon cannot change sizes, and it is not closer to Earth when it is rising. Although currently there is no agreed-upon explanation of the moon illusion, all the proposed theories appeal to various aspects of human vision to explain the illusion.[20] All this extra information is interesting, but it does not change the fact that you *perceive* the horizon moon to be larger than the zenith moon. That's why there is something to explain here in the first place. So, you cannot be wrong about your perceptions. And the same is true of all your conscious experiences: you cannot be wrong about their content. As philosophers put it: you are *incorrigible* about the contents of your consciousness.

So, we have direct, unmediated knowledge of our conscious contents. To get from here to genuine knowledge of the world, we only need two other ingredients: *confirming coherence* and *projection*.

Confirming coherence is the experience of other things that cohere with, make sense given, an original experience. This coherence tends to *mutually strengthen* the relevant set of experiences. It is relatively easy for this mutual strengthening to cross some threshold from mere belief to knowledge, and when it does, we *know*, rather than merely believe. There are several different kinds of coherence at work here. One kind is just persistence of the experience over the time you attend to it. Another kind is the experience's explanatory coherence with other experiences. A third kind is the experience's logical coherence with other experiences. And a fourth kind is the experience's coherence with your memory of other experiences you've had.

Projection is the perceptual process whereby your conscious experiences are experienced *as being about an external environment* – experiences are *projected* out into an external environment. A good argument that projection exists is to note that while you cannot be wrong about your experiences, you can be wrong that your external environment is the way you are experiencing it. So, there is a large epistemic distinction – not-possibly-wrong versus possibly-wrong – between your inner and outer realms. Consequently, the inner and outer realms cannot be the same. But 'wrong' here means that confirming coherence fails. Projection, then, is the process of capitalizing on this failure to project an external, real world. So, the separation of inner and outer realms is real (or experienced as real) and something we experience often. The moon illusion is just one example. The movie *The Matrix* is a well-told tale of another: Our lives strike us as real, lived in a real world; but this could all be completely wrong – we could be bodies confined to human-sized vats of goo forced to falsely experience the life we think we are living. (For more, see the Third War, Supplement 5.4: Problems with Consciousness.)

Usually, all these types of coherence and projection (plus others, probably) work together quickly. Let's call the whole process *projected confirming coherence* (*projected coherence*, for short). To slow down this process so we can see individual parts and inferences, let's pick an unusual example. Suppose you see a small strange object. You don't know what it is, but you already know you see it with your eyes. If the background makes sense to you – if you can categorize it, then projection already assures you that the strange object is *out there*. You aren't, probably, just imagining it or hallucinating.

Now suppose further that the strange object is sitting on a desk, off to the side of the computer sitting on the *same* desk (that it is the *same* desk is an instance of the first kind of coherence introduced above). With just this much projected coherence – desk, computer, to-the-side-of – you *know* there is something out there, but you are unable to categorize it. Further examination – getting closer to it, grabbing and turning it over in your hands – reveals that it is an oddly shaped coffee cup . . . which changes its colour as you move relative to it (the cup's glaze has been applied using lenticular printing – see, e.g. the Wikipedia page on such printing). Now you know what the object is. And you know this because all the information you gained perceiving and examining the cup coheres into one very plausible story: an oddly shaped coffee cup that changes colours as you move with respect to it.

This sort of example can be multiplied. But the point should be clear. Multiple kinds of perceptual information that forms the content of your consciousness, about which you cannot be wrong, projectively cohere into a story or an event about which you can be wrong.

Interestingly, failure to cohere also provides knowledge. The reason that we know that the moon illusion is an illusion is that we know the distance from Earth to the moon, and we know the shape of its orbit. The moon's orbit is elliptical, of course, and so not circular, so its distance does change somewhat, but not enough to explain the large size difference between the horizon moon and the zenith moon. And of course, the moon could not change sizes like a balloon losing and gaining air. Solid, massive moons don't behave like that. Hence, we have failure of coherence. The moon seems to change size somehow, but it can't. So, there must be some other kind of explanation. One involving the human vision system is a good place to look.

We conclude: projected coherence constitutes our knowledge of the 'world'. But consciousness is required for projected coherence. So conscious is required for knowledge. If the robot sharpener is to have any knowledge, then it must be conscious. And, crediting the robot with knowledge seems to be a very reasonable, even an ineluctably correct thing to do, given its abilities and internal computational processing. The nice thing here is, given projective coherence, consciousness, which one might be reluctant to impute to the robot, is naturally imputed because it is necessary for knowledge, which is much easier to credit the robot with.[21]

But knowledge will not exist as a solitary mental property. A sense of what matters must also be present.

Mattering Requires Consciousness

As argued in section 6, the robot pencil sharpener will have to have a complex inner mental life. Not, perhaps, as complex as yours, but perhaps nearly as complex as a dog's. One thing we did not discuss in section 6 is the notion of *mattering*. Things *matter* to you and to dogs and probably to all animals, certainly to all vertebrates. To work as well as it does, sharpening pencils will have to *matter* to the robot. It is not often noted, but *all* robots with the same level of autonomy as the robot sharpener – as a somewhat simple animal – will be successful only if their tasks and their environments matter to them. Yes, industrial robots, which do the same thing over and over and over, work well without, seemingly, anything mattering to them. But industrial robots have at best what we called monointelligence (see the Third War's Supplement 5.1): They can only do one or two things, and they are rarely autonomous. So, they are extremely unlikely to have knowledge and hence are unlikely to be or need to be conscious. By contrast, humans have *open-ended intelligence*: we can know about and figure out an open-ended class of things and problems.

Mattering is an extremely complex notion, involving at least consciousness, the self, values, worth, ethics, and points of view. It has not received the attention it deserves from either philosophers or cognitive scientists (but see e.g. Goldstein, 2017; O'Brien, 1996; and Nagel, 1979a). Very roughly, we define mattering thusly: *Something matters to a thinking being if that thing has worth to that being.* This 'definition' doesn't make much progress, relying as it does on the equally complicated notion of *worth*. But it will have to do. Of course, mattering is not binary, there are degrees of mattering: some things matter a lot, some matter only a little, and many things matter to an amount somewhere in-between these two. Also, what matters can change. Something can matter a lot at an early time in a person's life, say, and later, not matter very much at all.

Claim 1: Mattering is *felt*. One knows what matters to one by consciously introspecting. One has direct access to what matters to one (one can imagine that a person or being has to *discover* some of the things that matter it, but this discovery is just bringing what matters to consciousness). So, consciousness is required for mattering. This means both that one *knows* what *matters* to one, *and* it *matters* that one *knows* that one has such knowledge.

Claim 2: That they matter is what makes us want to do a good job at our various tasks.

Claim 3. It is plausible that with a complex, thinking being, an agent, mattering is required in order to prioritize tasks. Logical and physical relations also impact prioritization, of course (it's socks then shoes, not the reverse), but mattering plays an important role, too.

First Conclusion: For the robot sharpener to do a good job at sharpening pencils, sharpening pencils has to matter to it. This mattering will influence how it prioritizes its tasks and subtasks.

Claim 4: It is not possible that exactly one thing matters to a conscious thinking being. If anything matters to such a being, many things matter.

(Claims 1-4 seem obvious to us, and we hope they seem so to our readers; they are such basic claims that it is hard to see how to argue for them.)

There is one last consideration. Since your robot sharpener is conscious, you have to worry about it getting *bored*. Sharpening pencils is not exciting, at least after the first several hundred pencils. So, other things may come to matter to the robot, without you doing any extra programming at all – like the terrible way you leave your books all over the floor, making a navigation nightmare. Perhaps your robot will undertake, on its own, to put your books on a shelf.

Second Conclusion: To do a good job, several things (including perhaps new things) will have to matter to the robot pencil sharpener. Of course, all this mattering requires that the robot be conscious.

Conclusion to section 13

We now have two jobs for consciousness: knowing and mattering. These two properties seem crucial to having a robot sharpener with aboutness. And they also seem jointly *sufficient*. Armed with genuine knowledge and a sense of what matters, your robot *must* be conscious. Its mattering and knowing make its computations and data structures genuinely *about* objects and tasks it has in its environment, not just doing the tasks and finding the objects.

In fact, now, we see that in order to do the tasks and find the objects, the robot sharpener will *have* to be conscious. Let's now return to what we said at the end of section 12:

> So now we have a standoff. The 'aboutness-is-doing' crowd has no business assuming that the robot sharpener is not conscious – that it can work without being conscious. But the 'aboutness-is-consciousness' crowd has no business assuming that the robot sharpener needs consciousness to work, nor that AI in general needs conscious computers to finally produce a genuine intelligent, thinking machine.

The standoff is resolved. The aboutness-is-consciousness group wins. Any autonomous robot able to do complex tasks and navigate a complex environment will require both knowledge and a sense of what matters to it. These two also require consciousness.

14. The Vexed Task of Implementing Consciousness

The AI Wars concerning mental content were fought with almost no attention paid to conscious, real knowing, and mattering. Consciousness, as mentioned in section 10, was eschewed, ignored by most cognitive scientists of whatever type. In fact, as noted in section 12, the aboutness-is-consciousness group, who took both consciousness and aboutness seriously, was quite small, and contained no AI researchers. Yet, as we have seen, consciousness is likely crucial to building a genuinely intelligent machine.

So, how do we do it? Can we actually program consciousness? Can we program mattering and knowing?

No we cannot. For the detailed philosophical reasons why, see the Third War Supplement 5.4: Problems with Consciousness. And see Chapter 7. But we can glimpse the problem right here.

As we saw in section 10, consciousness is only a first-person phenomenon; there is no third-person, objective way to characterize consciousness. This is why there is only a weak, ostensive definition of consciousness. Remember, in section 10, we said:

[This] ostensive definition of consciousness, appealing to something we can only assume our readers have, will have to do, for there is no more robust scientific, third-person definition available.

And this, as we saw in section 10, is why there is so little productive, revelatory science done on consciousness. This first-person-only situation with consciousness is such a profound and unique roadblock to science, that distinguished philosophers have argued that consciousness is not a physical property of the universe (such a view is often called *dualism*, see Chalmers, 1996). Since scientific inquiry is barred from studying consciousness in any deep way, we cannot know what it is in any detail (see Dietrich and Hardcastle, 2004, for an extended argument for this). Such a thorough ignorance prevents us from implementing it.

There is, however, some good news, perhaps. Given consciousness's unique and strange status in our universe (as noted, it may not even be a physical property at all), perhaps consciousness is got for *free*. Though it is impossible to implement it directly, perhaps consciousness is a free side-effect of other implementations. So, it could well be that the robot pencil sharpener is conscious without us having to implement consciousness. Just implementing a robot that can interact robustly in its environment might be metaphysically sufficient for consciousness. So, we do not need to directly implement a conscious robot pencil sharpener, we need only to implement a robot that deftly sharpens pencils while not knocking over your coffee and we get consciousness for free.

We now get this interesting and surprising expression:

Aboutness-is-doing gives rise to Aboutness-is-consciousness.

Perhaps, even,

Aboutness-is-doing = Aboutness-is-consciousness

Let's call the first expression *DGRC* ('doing gives rise to consciousness') and the second *D=C* ('doing equals consciousness').

One very interesting conclusion that results from these expressions and this understanding of machine consciousness is this: If we implement the robot pencil sharpener, and it is conscious automatically, in virtue of our implementation, and if consciousness is crucial to robustly intelligent behaviour (section 13), then our robot sharpener might perform better than our implementation warrants, and better than our implementation can explain. And now of course, with consciousness, our robot sharpener has acquired moral standing. And we have to treat it well. See Chapters 8 and 9.

15. Conclusion to the Third War

We have seen in this chapter that the War Over Aboutness, over whether or not computers' internal data structures can be about things for the computers and not just for us users, was fought in a profound state of confusion. It does look like aboutness is consciousness, but consciousness, being such a troubled notion, was deliberately ignored both in the history of AI and, for the most part, now. Instead, notions of 'the right causal connection', 'the causal powers of brains', 'being embodied', 'being a dynamical system' were pursued. It all came to naught. The world has many, many robots today. But whether they are anything more than mere tools, like our refrigerators, is anyone's guess. Like so many wars, intellectual and armoured, the War Over Aboutness generated a lot of heat, but no light. To this day, neither AI researchers nor philosophers agree on what aboutness is or whether or not it is useful for intelligence. We have presented a case here that aboutness is consciousness and that it is required for truly intelligent robots. But given the troubled nature of consciousness, this could easily be perceived as just a continuing skirmish. (For another telling example of a continuing skirmish, see the Third War Supplement 5.5: A Recent Skirmish in the AI Wars.)

Third War Supplement 1

Examples of Unintelligent AI: Monointelligence is not real intelligence

To further support our claim that AI has not succeeded in building an intelligent machine, we briefly discuss several successful AI programs. None of these change our assessment of the unintelligence of AI machines.

AlphaGo and Friends

As we concluded in the main text, AlphaGo, once the world-champion Go player, was merely a monointelligent game player. It had nothing like human intelligence. But then AlphaGo Zero easily beat AlphaGo. Same conclusion. Then AlphaZero beat AlphaGo Zero at Go, and also, via learning, mastered chess — AlphaZero can now beat any human chess champion and all computer champions (Hutson, 2018) ... Same conclusion: impressive monointelligence, but hardly an AI worthy of the name.

Watson

Watson is an IBM supercomputer that uses AI machine learning techniques and other analytical algorithms to answer questions posed in natural language (English, for example). Watson is currently the reigning champion *Jeopardy!* player – *Jeopardy!* is a television game show. In 2011, Watson beat the two best *Jeopardy!*-playing humans: Brad Rutter and Ken Jennings.

Playing *Jeopardy!* was just a test environment for Watson – a flashy test environment, heavily covered by the media. But IBM is quite serious about using Watson in the world to help out humans. For example, IBM is attempting to deploy Watson in various spots in the medical field so that it can help out doctors and their patients. So far, these attempts have met with little success, but there may be hope on the horizon (see https://www.technologyreview.com/s/607965/a-reality-check-for-ibms-ai-ambitions/). However, regardless of how successful Watson is in medicine, it will still do *one thing*: answer medical questions. Doctors do this *and* change diapers, plan invasions, drive cars, design buildings, write sonnets, balance accounts, build shelters, set bones, comfort the dying, take orders, give orders, cooperate, act alone, solve equations, analyse new problems, pitch manure, program computers, cook meals, fight efficiently, and die gallantly ... and 'build' Watsons – all with a brain that weighs considerably less than Watson.

DeepStack

DeepStack is a game-playing computer. But is quite different from AlphaGo, AlphaZero and Watson (in its game-playing mode). All game playing programs work as well as they do partly because they all play a specific kind of game: games where all the players know all there is to know about the current status of the game. Furthermore, this fact is mutually known among the players: everyone knows everything relevant (except the answers, of course). This is usually put by saying that all the players have *perfect information* about the current game. So, computers played lousy poker because in poker, there is only *imperfect information*: one player does not know what cards the others have; there is bluffing, deception, and trying to figure out if the other players know what you are going to do. Poker is much more like real life, where important decisions have to be made knowing a tiny fraction of the relevant information. DeepStack plays poker, and is very good at it (see Moravčik et al. 2017). DeepStack specializes in heads-up no-limit Texas hold'em (though it can play several imperfect information games, p. 509). DeepStack has beaten eleven professional poker players.

Again, though, DeepStack is a gigantic *monointelligence*. It is nothing like the eleven professionals it beat: it is far inferior to them in terms of actually getting along in the world, in terms of having a *cognitive world* (see section 8). It is therefore far less intelligent than humans – in fact, it is not clear that it is even on the spectrum of kinds of natural intelligences. The average beetle on an ordinary day does far more than DeepStack could ever do.

Third War Supplement 2

Standard Answers to the Chinese Room Argument

The replies to this argument would fill a library of many rooms. AI-friendly philosophers and AI researchers laboured to refute it. Searle claims to have handled all alleged refutations (see his 1980, 1990, 1992). Here are some of the better-known replies, already discussed in Searle's original paper:

The System Reply

The System Reply states that while the Chinese symbols are meaningless to the person in the room, the person is only part of a *larger system* (including the rulebook, pencils, paper, and possibly other things, like the things keeping the person alive), and it is possible that that larger system has contentful thoughts.

The Robot Reply

The Robot Reply suggests making the room into the head of robot (a large robot), by attaching cameras and microphones for vision and hearing, and actuators to manipulate its environment, and having this robot interact with the group of Chinese historians. The Robot Reply is thus a specific instance of the System Reply, in that the thought is that maybe the larger system (the robot), of which the person is merely a part, has contentful thoughts.

The Brain Simulator Reply

The Brain Simulator Reply suggests that, at least in principle, the rules in the rulebook could form a simulation of what is going on inside the brain of a person who does speak Chinese, and whose understanding and intentionality with regard to those symbols presumably come from his brain. As such, the Brain Simulator Reply is once again a kind of System Reply: this time, the system of person and rulebook implement the working of a brain, thus potentially giving rise to the relevant intentionality that Searle claims is missing.

Third War Supplement 3

Semantics and Causation

The debate between Dretske and Rapaport, discussed in section 5 above, rests on a much deeper debate about the nature of semantics and hence aboutness itself. This deeper debate concerns the relation between semantics and *causation*. It continues to this day, and can seen as a continuation, in somewhat different terms, of the Third War.

Dretske's position follows from his causal theory of reference (Dretske, 1981): basically, that an internal state s of some system S refers to, if anything, the states of the external (to S) world that cause S to be in the particular state s. Your concept [Dog], for example, refers to dogs because dogs cause you – via complicated mechanisms described by disciplines such as psychophysics – to activate your dog concept [Dog] when you perceive a dog. For this to be the case, S clearly must be *situated* in some world, which we will call WS for 'S's world' or 'everything but S'. S must also be *embodied*, i.e. S must be implemented in some physical structure that has physical – hence ordinary causal – interactions with WS. Here 'situated' and 'embodied' are the architectural concepts encountered in Chapter 4.

This notion of situatedness as a requirement for a semantics driven by causation has been given a rigorous formal characterization by Barwise and Perry (1983) and extended by Barwise and Seligman (1997). This formalization of a causally driven semantics has had a large impact on distributed computing, machine learning, and robotics. It is hard, for example, not to regard the concepts learned by an ANN or other machine-learning algorithm as both caused by and referring to the training sets employed to 'teach' the algorithm each of the relevant concepts. It is hard to imagine the meaning, for the robot, of a robot's internal representations not having something to do with the robot's ability to manipulate the objects in its environment.

The causal antecedents of a symbol are also sometimes referred to as the 'ground' of that symbol. The 'symbol-grounding problem' is the problem of discovering (or rendering formally) the grounds for a given symbol (Harnad, 1990; Taddeo and Floridi, 2005). The symbol grounding problem is, it turns out, equivalent to the problem of identifying a quantum system given a finite description (Fields, 2014). This latter problem is unsolvable by finite means; hence the symbol grounding problem is similarly unsolvable.

Any causal notion of semantics is faced with a 'forward' problem analogous to the 'reverse' problem of the unsolvability of symbol grounding. This forward problem is that under very general assumptions (how general is still subject to debate), the states of physical systems are conditionally independent of the states of their environments. This conditional independence can be formulated using the concept of a 'Markov blanket' (see Clark, 2017 for an intuitive presentation and Kuchling et al., 2019 for full formal details). If s is conditionally independent of the state of WS, the state of WS cannot plausibly be regarded as either the cause or the meaning of s.

We are left, in this case, with something like Rapaport's position, which internalizes Wittgenstein's dictum that 'meaning is use' (Wittgenstein, 1953). 'Dog' refers to dogs not because dogs cause 'Dog' (or uses of 'Dog'), but merely because symbol users use 'Dog' (for whatever reason) to refer to dogs. It may seem difficult to see this as a deep understanding of the meaning of 'Dog'. What may, however, be a deep understanding is that all other attempts to explain what 'Dog' means have, at least so far, ended in failure.

Third War Supplement 4

Problems with Consciousness

A moment's reflection shows why consciousness was and remains ignored throughout AI and cognitive science: consciousness brings with it a host of extreme difficulties, while at the same time appearing to be irrelevant. One difficulty, as discussed above in section 10, is the lack of any scientifically useful, third-person definition of consciousness. This lack of a definition prevents us from directly implementing consciousness – the detailed nature of implementation requires that programmers have at least a definition, if not a full understanding, of what they are implementing. So barring a programming miracle of the kind discussed in section 14 (we get consciousness for free by implementing sensorimotor abilities and capacities), an implementation of consciousness eludes us. But there appears to be no downside to this, for perhaps the best explanation of AI and cognitive science ignoring consciousness is that all of cognition, all of thinking, as well as all of our moving around and doing things seems like it can go on perfectly well without consciousness. Consciousness, in short, seems to be an *afterthought*, literally: all of our important thinking and perceiving occurs and then (milliseconds later) we are conscious of what we are thinking about or what we are perceiving (for the central research on this topic, see Libet, 1985a, 1985b, 1987, 1989a, 1989b, 1991). A useful metaphor here is that our conscious lives often strike us as a movie we are watching. But the events in our lives will go on whether we watch or not: the movie plays even if no one is there to see it. 'Consciousness doesn't appear important for living and doing', is one strong way to put this idea (for more on consciousness being an afterthought, see Harnad, 1982.) Some in AI seemed to agree with this: McCarthy (1979), for example. Let's call this afterthought property of consciousness *passive consciousness*: consciousness is a kind of passive watching, and nothing more. (This of course clashes with our conclusions from section 14 – DGRC ('doing gives rise to consciousness') and D=C ('doing equals consciousness'). We return to this point at the end of this supplement.)

It should be stressed though, that consciousness is required for *emotions*. Emotions differ from thinking on precisely this point. Fear, for example, would be nonexistent if it were somehow unconscious. Perhaps the notion of unconscious fear is incoherent, like the notion of a four-sided triangle. Yes, something one is not conscious of might make one's heart speed up, as with fear, but the person in question would not attribute that speed-up to fear and would not in fact feel fearful. Fear, as with all the emotions, even those we don't have good names for, must be felt to exist. But, with their demotion of consciousness in general, AI scientists have usually dismissed emotions, too, as not required for intelligent thought. We return to this point below.

There is a philosophical argument that consciousness is not required for intelligent thought. Before we present this argument, though, it is crucial to note that all the arguments below require that we assume that the expressions derived in section 14, *DGRC* ('doing gives rise to consciousness') and *D=C* ('doing equals consciousness') are wrong. Why? Because if the expressions are right, then it is physically and psychologically impossible to implement an intelligent, *unconscious* being that successfully inhabits a cognitive world.

The main way philosophers attempt to show that thinking doesn't require consciousness is to invoke *zombies*. (A somewhat weaker argument for a similar conclusion is the inverted spectra argument. We present this argument, also.) Using the term from section 8 – *cognitive world* – zombie arguments show us that we do not need to be conscious of our cognitive world at all. An unsettling conclusion.

Thinking Without Consciousness: Zombies

Zombies raise serious problems for the study of consciousness and for the goal of building conscious machines. We are not interested in the Hollywood kind which run or stumble around biting or eating people, perhaps eating their brains, and turning the bitees into zombies too. We are interested only in the philosophical kind who look and act and behave just like you do, but who are not conscious at all. The easiest way to conjure a philosophical zombie in your mind's eye is to imagine an exact *physical replica* of you (an atom-for-atom replica, say), but without consciousness (see Chalmers, 1996). Call this your *zombie twin*.

Claim: Your zombie twin is not conscious. It behaves like you though, going to work or school, interacting with its family and friends, eating and drinking, laughing and crying, etc. But inside your zombie twin, there is nothing going on at all. Your zombie twin has *no experiences whatsoever*. Not a single momentary flash of light, ever.

Why believe that this mere physical replica could do anything other than just lay on the ground? Because your zombie twin can still process the energy impinging on its sense organs, its sensory neurons still fire, its brain still works (brain neurons also fire), and so do its muscle neurons, contracting its muscles. Shown a red apple, your zombie twin will say quite clearly that it sees a red apple. It has no experiences whatsoever, but light from the apple still enters your twin's eyes, then its retinas process this information and pass it onto its visual cortex and related parts of its brain, where the process gets more involved and more complicated. Finally, the highly processed information is then sent to the part of its brain responsible for speaking – its lips, tongue, and velum work together to say, 'I see a red apple.' In fact, your zombie twin might grasp the apple, take a bite and say: 'Yum!' There'd be no conscious taste of the apple whatsoever, but its behaviour would be just like yours. Using our Nagel characterization of consciousness (Nagel, 1974, and see section 10): There is nothing it is like to be a conscious person's zombie twin, just like there is nothing it is like to be a doorknob or a rock heating up in the sun – the rock doesn't feel warm, the doorknob doesn't experience rotating.

Chalmers, the champion of zombies in philosophy, put it this way (*functionalism* or the *functionalist approach* was defined in section 6):

> What is going on in my zombie twin? He is physically identical to me, and we may as well suppose he is embedded in an identical environment. He will certainly be identical to me *functionally*: he will be processing the same sort of information, reacting in a similar way to inputs, with his internal configurations being modified appropriately and with indistinguishable behavior resulting. He will be *psychologically* identical to me ... He will be perceiving the trees outside, in the functional sense, and tasting the chocolate, in the psychological sense. All of this follows logically from the fact he is physically identical to me by virtue of the functional analyses of psychological notions. He will even be 'conscious' in the functional sense ... he will be awake, able to report the contents of his internal states, able to focus attention in various places, and so on. It is just that none of this functioning will be accompanied by any real conscious experience. There will be no phenomenal feel. There is nothing it is like to be a zombie.
>
> 1996, p. 95

Now for the take-home message. Successfully imagining a Zombie twin severs any (intuitive) close tie between thinking and consciousness by showing us that thinking can occur without any consciousness at all. Consciousness isn't even an afterthought; it is not needed in any way for any aspect of thinking.

However, AI scientists weren't zombiephiles. When they eschewed consciousness, they were not up on the latest in zombie philosophy. No AI scientist was going around saying that zombies proved that she didn't need to worry about consciousness. Nevertheless, the underlying intuition that makes zombie twins seem possible and hence that consciousness was not necessary was widely held. The evidence for this claim is the rampant pragmatism of AI researchers, as exemplified by McCarthy (1979), and the outright ignoring by AI researchers of the very essence of consciousness: subjective experience. (For more on this, see Supplement 5.5, next.)

However, there was a philosophical argument much more well-known and much more public than the zombie argument, and it too suggested that

AI scientists could ignore consciousness. This is the inverted spectrum argument.

Thinking Without Consciousness: The Inverted Spectrum

Suppose that the spectrum of colours you see as you go about your life is not the spectrum your best friend sees. Your friend sees a spectrum inverted relative to yours. Here are the colours you see on the left and the colours you best friend sees on the right:

Red	Violet
Orange	Indigo
Yellow	Blue
Green	Green
Blue	Yellow
Indigo	Orange
Violet	Red

So, where you see a blue sky on a cloudless summer day, your friend sees a yellow sky. In fact, all things that look blue to you look yellow to your friend. But, and this is the key, your friend uses the word 'blue' to name the colour of the sky. Why? Because when your friend was learning English, someone pointed to the sky and said, 'That colour is blue.' So, your friend learned that the colour you would call yellow if you saw your friend's spectrum is named by the word 'blue'. Since this is *exactly* the way you learned that the sky is blue, you also use 'blue' to describe the colour of the sky. Your friend can also point out that blue berries aren't really blue; they are indigo. You agree. But where you see indigo blueberries, your friend sees orange berries. But as with the sky, your friend uses the word 'indigo' to name the colour he or she sees when looking at a blueberry.

Conclusion: there is no way for you to find out that your friend's colour spectrum is inverted relative to yours. What this means is that, when it comes to colour experiences, it doesn't matter what colour you see. You can get along perfectly well with an inverted spectrum. This in turn means, as we have stressed, that only *your* conscious colour experiences are accessible to you. There is no public way to get at your friend's conscious colour experiences, nor can anyone else get at yours. And this means that colour consciousness is irrelevant to dealing with colour in your daily life. You and your friend communicate perfectly well about the blue sky, all the while having radically different experiences of it. Since the actual experiences don't matter, consciousness appears be to an extraneous add-on.

So, we get this overall conclusion: Consciousness was widely ignored in AI because:

1   consciousness seemed passive;
2   spectrum inversion indicated that it doesn't really matter what colours a thinking being sees; and
3   there was an underlying, persistent intuition that it doesn't even matter *whether* a thinking being sees colors at all.

Please note: if our theory presented in section 13 is right, then the above is wrong. Zombies, though they seem possible, and are conceivable, are in fact not possible. And consciousness is *not* passive. (The status of spectrum inversion remains undetermined.) If section 13 is right, non-conscious, intelligent thinking is not possible.

Is the Robot Pencil Sharpener Conscious?

We now know the answer: *Who knows?* Even though, shown a working robot pencil sharpener, you'd find it impossible *not* to think of it as conscious, it might not be conscious. If the AI scientists are right, then it doesn't matter if it is conscious. If section 13 is right, then it must be conscious, though of course, we could never know directly. All this is rather depressing. How can something as important as consciousness is in our own individual lives be so elusive and gossamer?

An answer to this question is provided in Dietrich and Hardcastle (2004). Their answer is that the very nature of consciousness itself precludes studying it in any way, whether or not dualism is true (dualism is the thesis that consciousness is not a physical property of our universe – Chalmers, 1996, uses zombies and inverted spectra and other tools to argue for dualism – if dualism is true, the science of consciousness will be stymied). Dietrich and Gillies (2001), argue that the very nature of consciousness itself makes it seem like dualism is true, even if it isn't. So, for all we will ever know, consciousness might arise quite naturally, inevitably, and even necessarily, from the firing of neurons or from the processing of information in any way. The truth about consciousness will forever elude us. This is certainly a strange conclusion.

Third War Supplement 5

A Recent Skirmish in the AI Wars: Conscious Machines

In 2017, Dehaene, Lau, and Kouider published an important paper in *Science* on conscious machines (computers and robots). The paper covers a lot of ground, but their central points are that (1) currently, machines are not conscious, but (2) they would be conscious if we implemented two properties: *global availability* and *self-monitoring*. Information is globally available, according to Dehaene et al., if it is available for recall, for directing action, and for speaking about, among other things (p. 486). Self-monitoring is the process whereby a machine collects data about its own processes (including collecting data about its own processes) in order to learn what it is doing and how well it is doing it. With this knowledge it can update databases, change goals, directions, revise plans, etc.

Global availability and self-monitoring are obviously very useful and building machines or robots with these two capabilities is a worthy research project. But they have nothing to do with consciousness. Here, again, is the 'definition' of consciousness from Dietrich and Hardcastle, 2004:

> Consciousness is the way the world seems to us, the way we experience it, feel it. Taste an onion, see a rainbow, smell a dead skunk on a hot summer's day, stub your toe on the foot of the bed frame at 4:00 A.M., hear your dog breathe or a baby gurgle and coo. These are experiences, bits of our phenomenology, and it is these experiences that somehow give us our subjective point of view ... We have experiences because we are conscious. Or, rather, our having them constitutes our being conscious. Being conscious is what makes it fun or horrible or merely boring to be a human. Using the phrase that Thomas Nagel (1974) made famous, we can say that a being is conscious if *there is something it is like to be that being.*
>
> Dietrich and Hardcastle, 2004, p. 5, emphasis added

Being able to taste an onion will not result from implementing global availability and self-monitoring. All computers do at least a bit of self-monitoring already, say, to efficiently allocate random-access memory. And much information on a computer is globally available to that computer. Does this really make ordinary, garden-variety computers conscious? It seems unlikely. (But who really knows?)

Yes, a consciousness machine might have a robust kind of global availability and self-monitoring, but it does not go the other way: global availability and self-monitoring do not cause or create consciousness.

Immediately, several neuroscientists and philosophers objected, saying:

> To determine whether machines are conscious, we must ask whether they have subjective experiences: Do machines consciously perceive and sense colors, sounds, and smells? Do they feel emotions? Unfortunately, Dehaene et al. relegate this issue to the final paragraph of their Review, dismissing it as a philosophical question 'beyond the scope of the present paper.' Instead, they ask whether machines 'mimic' consciousness by exhibiting the global availability of information (the ability to select, access, and report information) and metacognition (the capacity for self-monitoring and confidence estimation). Questions concerning to what extent machines have these capacities are interesting, but neither capacity is necessary or sufficient for subjective experience ... Furthermore, Dehaene et al.'s emphasis on metacognition and global broadcasting presumes that the prefrontal cortex is the home of consciousness, which remains a matter of debate ...
>
> Carter et al., 2018

So, a computer with global availability and self-monitoring might *give the appearance of being consciousness*, but it is not very plausible that such a machine would thereby *be* consciousness. Still ... if either 'doing gives rise to consciousness' or 'doing equals consciousness' is true then appearing to be conscious might be the right approach after all.

This recent skirmish embodies the history of AI's ignoring consciousness and its attempt to achieve aboutness via simulating aboutness. The heyday of this sort of the debate is passed, but it is not dead. Consciousness and aboutness are still of supreme importance, whatever they are.

Third War Supplement 6

The Chinese Room and the Fantastic Voyage Variation

A good way to see that the Chinese Room Argument does something illicit is to consider the *Fantastic Voyage Variation* (Dietrich, 2014). Suppose that you are in command of an army – a very special army. This army has been shrunk down to molecular size and inserted in an inert brain, called Brain X.

Your mission is to reactivate this brain *by hand*. So you have to: push the neurotransmitters around between neurons and their synapses to handle (by hand) the build-up of neurochemical electrical charges, to check threshold levels, to fire (by hand) the electrochemical neuro-mechanism that sends the electrical signal down the axon, and to do everything else that needs doing in a brain including making sure the neurons get nutrients (e.g. glucose), making sure the neurons excrete waste and moving that waste away, getting rid of dead neurons (a human brain loses thousands every day), making new neurons (a somewhat rare event) and new synaptic connections (less rare), making sure enough oxygen is present, taking in electrical firings from the sense organs, sending electrical firings out to the muscles, . . . and so on and so forth. In short, you and your army are going to be what makes Brain X work. It's a highly unusual job, but it is your job, so naturally you do the very best you can.

Due to your prowess as a commander, your homunculi army soon has Brain X up and running perfectly. Since you are a hands-on commander, you have detailed knowledge of what your army is doing at any given time. But if you are asked, 'What is Brain X thinking about?', you will have to answer, 'I have no idea. All I know is that such and such neurochemical supply is low and more is being made, that neurons 12,009,798,444 through 12,009,800,444 have died and are scheduled to be removed, that a certain collection of neurons in the prefrontal cortex are firing in this way and at this frequency . . .'

The Fantastic Voyage Variation (named after the 1966 science fiction film wherein scientists are shrunk down to microscopic size in order to repair a man's brain) is a close analogue to Searle's Chinese Room Argument. But here, it's not computation that is the target but actual detailed brain and neuron functioning. The Variation shows that semantic content is missing *at the level of the functioning neuron*. But we *know* that neural functioning *is* the basis of our contentful thought. (If this is false, then the most robust kind of dualism is true, and hence studying brains borders on being a waste of time.) So, the fact that content disappears at micro-levels is the relevant phenomenon, not computation, as Searle supposed. Computing turns out to be a red-herring.

Of course, it is a gift to the Variation to grant that it establishes its conclusion. The Variation is, like the Chinese Room Argument, little more than an intuition pump. Perhaps content doesn't really vanish at micro-levels; perhaps it only seems to because the intuition pump places fully conscious, intelligent beings at the lower levels. The semantics they are conscious of might not be the right semantics.

If content does, in fact, vanish, though, and the Variation is right, then there is something deep going on. This last claim is made all the more pointed by noting that even in naturally intelligent systems, like humans and snails, the nature of mental content remains mysterious – this explains why there is no agreement on what's wrong with Searle's Chinese Room Argument. Neither cognitive psychologists nor neuroscientists have a theory of mental content that works. Many scientists have offered their individual theories, but none have garnered anything resembling a favourable consensus. For completeness's sake, computer scientists study semantics under the guise of programming language semantics, which links with other fields such as program verification, compiler theory, category theory and language design. It is clear to no one how the results from these fields help, or even could help, the goal of figuring out mental semantics in living, thinking beings.

The Fourth War:
Rationality, Relevance, and
Creativity

6

What is Relevant to What?
The Frame Problem

1. Introduction: Why are Humans so Smart?

We all want to know where we fit into the universal scheme of things. Connecting fully to this desire and focusing on what seems to be our unique human intelligence, Aristotle defined us humans as *rational animals*. The other animals were, therefore, not rational, according to him (and nearly everyone else at the time). 'The animals other than [humans] live by appearances and memories, and have but little of connected experience; but the human race lives also by art and reasonings' (Aristotle, Book 1, Part 1, *Metaphysics* – Aristotle intends to include only rational reasonings; for him, there was no other kind of reasoning). Today, we know this is false, Aristotle was wrong: all animals are rational because all achieve their goals (or try to) using appropriate means. Sometimes the goals are only to eat, mate, and avoid predators, but these are, of course, worthy goals. Still, Aristotle did have an inkling of something important but impossible for him to articulate: *Humans are a species of African ape, but we also deserve our own phylum.*[1] We humans are all basically chimps, but we are also unique. It is plausible that chimps think of the Moon as some sort of light in the sky. But we know why it is a light in the sky and more – we have been there and back in rocket ships. This is what Aristotle was pointing at.

So the 'humans are cognitively special' idea has been with us for a long while, in one form or another. The frame problem begins here. It is the problem of what makes us cognitively special – of what makes us so smart. It is, therefore, a problem concerning rationality, or good reasoning. But where this good reasoning is applied in the frame problem is a bit surprising. The problem's central issue involves the notion of *change*: all living things live in changing environments; and among thinking things, some of the most important changes that they must deal with are changes they themselves bring about – moving from place to place, for example. However, more than change is involved. To successfully handle any change, thinking things must *update* what they believe after experiencing a change. Perhaps the major difficulty wrought by the frame problem is questioning how this updating should be carried out. If you put a pot on the stove to boil water, you need be aware that the lid, including its handle, is likely to get hot; so you can't just grab the lid's handle with your fingers to check on the boiling water or to put in, say, some rice or pasta. Putting a pot on the stove to boil water is a change, and this change requires updating one's belief about the temperature of the lid and its handle (however, the temperature of the pot itself quite naturally gets updated in your mind – you would never just grab the pot between your two hands). Failure to so update the temperature of the lid will result in burned fingers. This can be put in terms of *relevance*: the temperature of the pot is relevant to the temperature of the lid, so if the former changes, so does the latter. But this example barely scratches the surface.

The frame problem is a complicated problem with a tortured history. And it was the topic of the Fourth War. It is not possible to define the frame problem succinctly and easily. In fact, this was a big part of what the Fourth War was about.

The frame problem began life in an infamous 1969 paper by John McCarthy and Patrick Hayes ('Some Philosophical Problems From the Standpoint of Artificial Intelligence'). In this paper, McCarthy and Hayes introduced the term 'frame problem' to denote a seemingly narrow logic problem that arose, innocently enough, while the two worked to develop a logic for modelling good reasoning. But just nine years later, in 1978, the frame problem had become 'an abstract *epistemological* problem' (a problem about what we can know – see Dennett, 1978, p. 125, emphasis in original). Then, in 1987, the philosopher Jerry Fodor equated the frame problem with 'the problem of how the cognitive mind works' (Fodor, 1987, p. 148). He then claimed that understanding how the mind works requires unravelling the nature of inductive relevance and rationality (ibid.). So, solving the frame problem requires figuring out inductive relevance and rationality. (Inductive relevance is a technical term denoting how we figure out what information is relevant to proving a given scientific, or even an everyday, hypothesis.) From here, the frame problem continued to grow and expand until it covered vast areas of metaphysics, epistemology, philosophy of science, and philosophy of mind. By then, of course, the frame problem was revealed to be a serious and deep philosophical problem ... and therefore completely intractable.[2] Let's investigate what happened.

2. AI's Failure: Predicting it versus Explaining it

From the late 1950s to sometime around the turn of the millennium, AI scientists aimed to render humanity's cognitive specialness in silicon. Clearly, they did not succeed. The impressive success of the *monointelligences*, like Watson, AlphaGo, AlphaZero, DeepStack, and so on, do not seem to be bringing us any closer to machines with human-level intelligence – precisely because humans are not monointelligences, but *open-ended intelligences* (see the Third War, Chapter 5, on semantics and aboutness for a discussion of monointelligences and open-ended intelligence, and also Supplement 5.1). Something vitally important about human intelligence has been missed (and we'd be bizarrely lucky if it turned out to be only one thing).

It is a matter of historical record that the major philosophical objections that fuelled the AI Wars were *in-principle* objections to AI. These objections all had this form:

AI *will* fail because X.

Note the use of the future tense. Here, X ranges from 'syntactic digital computation is the wrong process', to 'symbol or representation manipulation is the wrong architecture', to 'computers lack aboutness'. All of these objections were driven by philosophical intuitions that computers necessarily lacked something – some *spark* – crucial to human-level intelligence. These attacks, as we have discussed in the first chapter of this book, ranged in years from 1961, when J. R. Lucas published 'Minds, Machines, and Gödel', to the long years of fighting after Searle published his attack arguing that machines lacked aboutness (published in 1980), to the battles over the frame problem, the nature of intelligent computation (assuming this phrase is not an oxymoron), and the fight over which architecture could actually realize intelligent computation – all of which lasted at least up through the turn of the millennium. During that whole time, say from 1961 to 2000, AI seemed to many philosophers like a research program that was ultimately doomed to failure. AI was impossible. What philosophers were doing was predicting *why AI was going to fail*. Future tense.

The frame problem is the one philosophical objection to AI that has been used to explain *why AI has failed thus far*. Past perfect tense. In 1987, the philosopher Jerry Fodor said:

> We can do science perfectly well without having a formal theory of [which ideas or events are relevant to each other]; which is to say that we can do science perfectly well without solving the frame problem. That's because doing science doesn't require having *mechanical* scientists; we have *us* instead. But we can't do AI perfectly well without having mechanical intelligence; doing AI perfectly well just *is* having mechanical intelligence. So we can't do AI without solving the frame problem. But *we don't know how* to solve the frame problem. That in a nutshell is why, though science works, AI doesn't.
>
> Fodor, 1987, p. 148, emphases in original

AI doesn't work, according to Fodor – at least it hadn't worked up to 1987 (and we've argued that it hasn't worked up to now – unless you count monointelligences; see Supplement 5.1). Fodor was using the frame problem to *explain why* AI hadn't worked, as of 1987. It is clear that, for Fodor, being mechanical carries serious liabilities. However, he is not denying that we can build mechanical scientists, and even mechanical open-ended intelligences, but he is denying that AI has produced any *reason* to think that mechanical open-ended intelligences are implementable – any reason beyond just faith.

3. How the Frame Problem Got Its Name: The Dream of Mechanization.

In order to fully understand the frame problem, we must give the history of how it got its name. As we said above, part of the war over the frame problem was over its name and which problem was picked out via that name. The other part of the war was over who got to use its name. Often, AI researchers and their friends use the story that *the AI* frame problem was really just a technical problem in the logic of change, first reported by McCarthy and Hayes in 1969, and that the problem has been more or less successfully solved (for the 'more or less', see below). So, today, there is no frame problem. These same researchers then go on to say that the infamous 'frame problem' is just some supposedly 'supercharged' problem dreamed up by pesky philosophers, and that the supercharged frame problem has nothing to do with the real AI frame problem. This 'two frame problems' narrative can be easily seen in such central works as Shanahan's discussion of the problem in the *Stanford Encyclopedia of Philosophy* (Shanahan, 2016). However, this narrative draws a distinction where none should be. On this point, the philosophers like Fodor seem to have been right. We here explore the 'two frame problems' narrative, and remove any alleged distinction that funds it. First, however, a preliminary issue: how logic, the dream of mechanization, rationality, and computers all came together.

In the mid-to-late 1800s and early 1900s, after a couple of millennia of mostly being ignored, logic was again receiving close attention from philosophers and mathematicians. During this time, logic grew and flourished. And during this time, logic was construed as a formalization of human thinking. Rationality, it was thought, was embodied in logic, or least it could be. A book title reveals this connection: In 1854, the mathematician George Boole published his famous book, *An Investigation of the Laws of Thought on Which are Founded the Mathematical Theories of Logic and Probabilities*.

At same time, the ancient dream of mechanization had taken on a new lustre of possibility.[3] A thorough history of this dream needs to be written, a history that captures both the profound hope associated with this dream as well as the hatred of this dream by all those who fear mechanization down to their innermost souls (Dietrich, 1995). For now, we just note that, in the seventeenth century, the philosopher and mathematician, Gottfried Leibniz, developed a primitive mechanical calculator, and then said, exemplifying the hope of mechanization:

> ... if controversies were to arise, there would be no more need of disputation between two philosophers than between two calculators. For it would suffice for them to take their pencils in their hands and to sit down at the abacus, and say to each other (and if they so wish also to a friend called to help): Let us calculate.
>
> Leibniz, 1685

But it was the work of the philosopher Gottlob Frege that began realizing the hope of mechanization, for it was Frege's work that was crucial to bringing logic and computation together. His well-known work, *Begriffsschrift, eine der arithmetischen nachgebildete Formelsprache des reinen Denkens [Concept-Script: A Formal Language for Pure Thought Modelled on that of Arithmetic]* (1879), aimed to get rid of the imprecision and ambiguity that (allegedly) bedevils ordinary language by capturing 'pure thought' free from bells and whistles and modelled on arithmetic (see van Heijenoort, 1967, ch. 1). The connection with arithmetic is crucial, for the thing about arithmetic is *all of its calculations are algorithms.* An algorithm is a series of simple, unambiguous steps that produce something (output) from some initial thing or things (input). The standard method of adding is a classic algorithm. So is the standard method of multiplying. And so forth. (See Chapter 3, section 3.1: the greatest common divisor algorithm.)

After Frege, logic and computation became best friends forever, both developing in sophistication and depth and both contributing to the other. It was a match made in heaven, at least Leibniz's heaven. Logic is mechanizable. It is something for which algorithms can be developed. And logic captures, or can capture, our rational thought (it was believed). So, our thought, our rational thinking, is mechanizable. This great early twentieth century insight was just waiting for the invention of the computer by Alan Turing and others. A computer is the mechanism everyone was looking for. It is thus no wonder that AI got started as soon as computers were available for more than calculating for the military. The year was 1956; the program (the *first* AI program) was Logic Theorist, developed by Allen Newell, Herbert Simon, and Cliff Shaw. Logic Theorist was touted as using some of the same problem solving skills as humans, and it even proved some logic theorems from Russell's and Whitehead's magisterial *Principia Mathematica* – a book that finally grounded mathematics (number theory, specifically) in logic. After Logic Theorist, the sky was the limit . . . seemingly.

4. How the Frame Problem Got Its Name: Logic and AI.

So it is no accident or mere coincidence that AI researchers (who were computer scientists, after all) used logic to develop their theories of mechanical, rational thinking. But by its very nature, any project to build thinking machines must come to terms with some very difficult philosophy problems. Logic, thinking, rationality, mechanization, and philosophy all came together in 1969, when McCarthy and Hayes published their paper 'Some Philosophical Problems from the Standpoint of Artificial Intelligence'. Here are their first two paragraphs:

> A computer program capable of acting intelligently in the world must have a general representation of the world in terms of which its inputs are interpreted. Designing such a program requires commitments about what knowledge is and how it is obtained. Thus, some of the major traditional problems of philosophy arise in artificial intelligence.
>
> More specifically, we want a computer program that decides what to do by inferring in a formal language that a certain strategy will achieve its assigned goal. This requires formalizing concepts of causality, ability, and knowledge. Such formalisms are also considered in philosophical logic.

And not just in philosophical logic, but in philosophy, in general.

The stage is now set. So, how did the frame problem get its name, and why does it matter?

Time, Situations, and Fluents

To use logic to develop a reasoning engine, the logic has to be able to handle change. Usually, classical logic is not used for such a thing because it doesn't have the required resources. (Classical logic is the best understood and most widely used logic throughout mathematics and science. Its roots date back at least to Aristotle.) Classical logic is used to handle possible truths (also called *conditional truths*) that have the air of the subjunctive about them. This is most easily seen in the conditional, 'If P then Q', where P and Q are sentences. Here is an example from mathematics: 'If triangle ABC is a right triangle, then the square on its hypotenuse is equal to the sum of the squares of the other two sides.' ABC need not be a right triangle, but *if* it is, then the Pythagorean Theorem is true of it. Note, if ABC is a right triangle, then it is a right triangle for eternity. And hence the Pythagorean Theorem is true of it eternally. These properties or conditions will never change.

But classical logic can handle ordinary, quotidian reasoning, too. Here, the conditions are true only transitorily. For an example, consider the conditional: 'If it is raining, then the streets are wet.' Again, classical logic is not being used to handle change, but is rather being used to handle possible, transitory truth: '*If* it is raining, *then* the streets are wet.' This means, roughly, 'It is possible that it is raining (or 'it could be raining'), and if it is, then the rain is making the streets wet.' Given what we know about rain, it will rain only for a while. Note again, though, that classical logic is not being used to handle change, it is being used to reason with sentences that may be true, could be true, but which, if true, will probably not remain true. Nothing about change is captured by classical logic, rather the relevant reasoning is restricted to just those times and places where the relevant sentences are true – If it is raining now, where you are, then the streets are wet now, where you are.[4]

Classical logicians are fully aware of all of this. So, to handle change, logicians usually introduce some notion of *time*. This is what McCarthy and Hayes did. Some sentences are true at a certain time, others are false; but at a later time, which sentences are true or false *changes*.

The technical device McCarthy and Hayes used is a *situation* (for this subsection, all citations are to their 1969). Their definition of a situation: 'A situation, *s*, is a complete state of the universe at an instant of time' (section 3.1). They note that it is impossible to describe a situation completely, so they opt for (very) partial descriptions, which they call *fluents*. A fluent is a sentence *in logic* that is a condition, property, or state of affairs that can change over time. Fluents describe changing parts of situations.

Here is an example. Suppose that there is a robot pencil sharpener that moves around in a large lab locating and sharpening dull pencils, and then putting them back on the relevant table, all without getting in anyone's way or knocking over a test tube containing a dangerous pathogen (see the Third War for a description of this robot). Assume that the robot has found a dull pencil and is sharpening it. Assume that the robot's name for itself is Rob. Then the robot could represent this state of affairs using this fluent:

(A) Is-sharpening (Rob, dull pencil-293, s_1).

In English this says: '[I,] Rob, am sharpening the dull pencil numbered 293 in situation s_1.' (This isn't quite right because of the indexical 'I', but it is close enough for our purposes.) The situation, s_1, implicitly fixes a time (which in this case would be called colloquially 'now'). After the pencil is sharpened, the situation s_1 changes, and this new fluent is needed.

(B) Is-sharpened (pencil-293, s_2).

The situation has changed from s_1 to s_2, and the time, accordingly, has progressed to a new 'now'. Note that this second fluent, (B), would be false in the situation s_1. And (A) is false in s_2. So, change has been captured, in a way. Fluents can get much more complicated, but the idea remains the same.

Fluents were logical sentences used to state facts, beliefs, causality, plans, goals, actions, reasoning about the past (a past situation), present (the current one), and future (the next situation), the order of knowledge (one has to know the combination of a safe before one can open it), and a large number of other things that seem to constitute humans' mental lives. Since fluents referred to situations, and situations captured change (the move from one situation to the next represented change), fluents could represent change. That was the idea anyway.

But situations and fluents caused a problem – the frame problem.

Suppose in a situation, Rob the robot is sharpening a pencil. This knowledge would be represented by a fluent. But then Rob finishes sharpening the pencil. This requires a change of situations from one containing fluents about sharpening to one containing fluents about having finished sharpening. Question: does Rob still have the pencil?

Humans are able to answer this question quickly. If we press you to answer it, you are likely to get a bit of annoyed. *Of course* Rob still has the pencil, this is just *common sense*. Rob just finished sharpening the pencil and so still has it. Rob will return it to the appropriate desk (the one where Rob found said pencil) in a moment.

Unfortunately, classical logic augmented with situations and fluents cannot prove that Rob still has the pencil – not without adding in some assumptions or hypotheses to the effect that if X has a pencil and sharpens it, then when the pencil is sharpened, X still has the pencil.[5] But McCarthy and Hayes point out that this doesn't solve the problem. There are *millions* of things (at least!) that don't change when Rob finishes sharpening a pencil – for example: Does the pencil still have an eraser? Is the pencil lead the same colour as it was? Is the pencil the same colour? Does it weigh pretty much the same? Is it still on planet Earth? Is Antarctica still cold? ... and so on and so forth. Only a tiny fraction of these actually need to be proved (e.g. that Rob should return the pencil now – *Not*, note, that Rob still has the pencil – that should be obvious). So, upon finishing sharpening the pencil, some very small number of things Rob knows or believes need to be updated (Rob should return the pencil), and the rest of Rob's beliefs can be ignored, like the fact that it is still cold in Antarctica. But which beliefs are 'the rest'? To figure out which beliefs don't need to be updated, *all* of Rob's beliefs will have to be checked. This will take a long time. Worse: most of Rob's beliefs are *irrelevant* to sharpening pencil number 293, so most of this checking would be a waste of time. What to do? *This* is the frame problem. Simply put, the frame problem is the problem of blocking the vast number of inferences and non-updates about what has *not* changed as the result of doing some action *A* while allowing the small number of inferences and updates about what has changed as a result of *A*. McCarthy and Hayes say:

> In the last section of part 3, in proving that one person could get into conversation [over the phone] with another, we were obliged to add the hypothesis that if a person has a telephone he still has it after looking up a number in the telephone book. If we had a number of actions to be performed in sequence we would have quite a number of conditions to write down that certain actions do not change the values of certain fluents. In fact with *n* actions and *m* fluents we might have to write down *mn* such conditions.
>
> We see two ways out of this difficulty. The first is to introduce the notion of [a] frame ... A number of fluents are declared as attached to the frame and the effect of an action is described by telling which fluents are changed, all others [attached to a different frame] being presumed unchanged.[6]

Sadly, McCarthy and Hayes never say in detail what a frame is, nor how a frame is related to a situation. But we can conclude from their 1969 paper and other relevant papers, that a frame is something like a subset of a situation; a frame is a part of a situation. Only the *relevant* fluents are attached to a given frame. So, when a change occurs, relative to a frame, only a small number of fluents need to be examined for updating. All the other frames (and their fluents) in that given situation remain unchanged and can be safely ignored.

And that is how the frame problem got its name.

5. The Frame Problem Comes into Focus.

One makes a change, say, finishes a cup of coffee. What beliefs and knowledge have to be updated as a result? Well, the coffee cup is now empty. So, there's no point in trying to drink from it. The cup will cool off, but is this worth an update? Is one now married or newly single? Is the moon still in its orbit? Is there still a place called Antarctica? One can easily write down hundreds of things that not only do not need updating, but which are obviously irrelevant to the coffee cup now being empty.

Hayes gives a nice definition of the frame problem (1987). He asks us to consider a case where someone goes through a door from room 1 to room 2. Hayes says that we want to be able to prove that when an agent (a thinking thing) goes from room 1 to room 2, then the agent is *in* room 2. (Again, this might seem obvious, and it is to our human readers, but not to a computer – to get it to be obvious was (and still is) one of the goals of AI). To get this conclusion that going from room 1 to room 2 puts one *in* room 2, we need *an axiom* to this effect. No problem. We simply add it in. Hayes then says:

> But here at last is the frame problem. With axioms [like the one we are adding about changing rooms], it is possible to infer what the immediate consequences of actions are. But what about the immediate non-consequences? When I go through a door, my position changes. But the color of my hair, and the positions of the cars in the streets, and the place my granny is sitting, don't change. In fact, most of the world carries on in just the same way that it did before . . . But since many of these things CAN change, they are described in our vocabulary as being relative to the time-instant [of changing rooms], so we cannot directly infer, as a matter of logic, that they are true [after I change rooms] just because they were [before I changed rooms]: This needs to be stated somehow in our axioms.

> p. 125

Then Hayes, continuing, points out:

> In this ontology, whenever something MIGHT change from one moment to another, we have to find some way of stating that it DOESN'T change whenever ANYTHING changes. And this seems silly, since almost all changes in fact change very little of the world. One feels that there should be some economical and principled way of succinctly saying what changes an action makes, without having to explicitly list all the things it doesn't change as well; yet there doesn't seem to be another way to do it. *That* is the frame problem.

> p. 125, emphasis in original

Given any change in some thinking thing's environment, only a tiny fraction of things that the thinking thing believes need to be updated. But *all* of what it believes will have to be *checked* to see whether or not any specific belief needs updating. This is far too much checking.

We can improve on Hayes's definition a bit. Suppose you are trying to write a letter to someone in a word-processor called the FP-Word-Processor. Your letter, as is usual, is stored as some file. But suppose that every time you typed a single key in the FP-Word-Processor, it signalled your computer to check its entire store of information on all its drives to see if anything besides the actual file you are working on needs updating. The file, of course, needs updating, but your list of favourite cake recipes does not, neither do the pictures of your cat. In fact, most items on your computer do *not* need updating. And worse, it is a waste of time to see if they need updating or not. Using the FP-Word-Processor, it would take you hours to write a simple, short article that should take only a few minutes. The FP-Word-Processor is a disaster.

A real word-processor limits what it 'thinks' needs updating to just the file you are working on (and possibly some supporting files like backup files). This keeps your real word-processor zipping along as you type sixty words per minute.

Such engineered, enforced limiting works because by far most of the changes made to a word-processing file require only changing that file (and supporting files). Hence all the other changes that might possibly need to be made can simply be ignored – letting you, the user, worry about distant changes that might need to be made given that you are writing the letter.

What Hayes wants, what all frame problem AI researchers want, is some *stability principle* or some *principle of inertia*, like the engineered, enforced limiting of a standard word-processor, which works *in all cases and for all tasks no matter how complicated.* AI researchers want some universal principle that captures the fact that most things not only don't change relative to a given change, but even worrying about any but a tiny fraction of these other things is a waste of time.

Specifically, what everyone wants is a general principle that looks like this:

> If a fluent *f* is normally or usually true in a given situation, and if, after a change of situation, the agent or thinking thing is not directly confronted with anything contradicting *f*, then the agent should assume that *f* is still probably true.[7]

But even this is not quite right. Given any change, we don't want to *explicitly assume* that most other things remain unchanged. We don't want to waste even one second thinking about those other things. We just want to proceed onwards, updating the obvious things that need updating. Unfortunately, no such general principle has been found – as both Hayes and Fodor note. What McCarthy and Hayes really did is trip over a heretofore underappreciated philosophy problem – a problem about how the world works and what we can know about it – a problem about the metaphysics of the world, and the associated epistemology of cognizing beings. Without AI, we might never have discovered how serious this problem is. And we would have never learned the deep truth about how humans 'solve' the frame problem (see section 8, below). The problem referred to here is *not* a problem limited to logic; rather it is the frame problem in full battle regalia: how do we algorithmically and efficiently determine what information is *relevant* to a given change? More simply: what is informational relevance and how do we determine it?

6. The Frame Problem Metastasizes

There are several technical issues surrounding what is known as the 'logic version' of the frame problem. In sum, what happened is that classical logic had to be abandoned and several new, unusual, nonclassical logics had to be investigated and deployed. (Good introductions to this are in Shanahan, 1997 and 2016.) Mostly, the accepted view now is that the logic version of the frame problem has been *more or less* solved. Shanahan (2016) says:

> . . . [A] number of solutions to the technical frame problem [the logic version] now exist that are adequate for logic-based AI research. Although improvements and extensions continue to be found, it is fair to say that the dust has settled, and that the frame problem, in its technical guise, is *more-or-less* solved.

> Emphasis added

So, this apparently leaves us with the philosophy version. But, as we stated in section 3, above, this two frame problems narrative draws a distinction where there shouldn't be one. The suggestion that the logic version of the frame problem was somehow different from the philosophy version allows AI researchers to paint philosophers as slightly hysterical in their concern over the frame problem, that the philosophers went a bit overboard when working on it. This is part of the reason for the two-version narrative: AI researchers simply denied that the philosophy version was connected to the 'real' frame problem. Hayes himself throws down the gauntlet when he says:

> The term 'frame problem' is due to John McCarthy, and was introduced in McCarthy and Hayes (1969). It is generally used within the AI field in something close to its original meaning. Others, however, *especially*

philosophers, sometimes interpret the term in different ways ... In this short paper I will try to state clearly and informally what the frame problem is, distinguish it from other problems with which it is often confused, briefly survey the currently available partial solutions, and respond to some of the *sillier misunderstandings*.

1987, p. 123; emphasis added; as Hayes's paper
proceeds, the language heats up

Sillier misunderstandings ... Hayes means, of course, *philosophical misunderstandings*. Well, let's face it, no one really likes philosophers. Athens was the Platonic (*ahem!*) ideal of this – the city officials gave Socrates a choice: leave or die. They did this because he kept asking embarrassing questions. (Socrates loved interacting with people, so leaving was not an option for him.) Asking embarrassing questions is every philosopher's job. Here are some embarrassing questions, related to the frame problem.

Recall Hayes's stated intuition, quoted in section 5, above: 'One feels that there should be some *economical* and *principled* way of succinctly saying what changes an action makes, without having to explicitly list all the things it doesn't change as well; *yet there doesn't seem to be another way to do it* (emphases added).

Why? Why is listing the memory or database items that don't need to be changed the only solution? Such listing is not principled and it is profoundly non-economical. So then why is a better way so hard to find? For example, why don't the things that should not even be considered for updating rest quietly in some undisturbed part of the database or memory? Are we stuck with an inelegant solution of merely managing long lists of things that don't change?

One can imagine Hayes, a champion of logic, suggesting that logic (classical logic, perhaps, or a nonclassical one) will provide the required *stability principle* or *principle of inertia*. But classical logic cannot even capture the simplest semantic relevance. The following conditional is *true* in classical logic: 'If President Obama is a prime number, then $2 + 3 = 19$.' Proponents of some nonclassical logics suggest they can rule out this conditional (see Beall et al. 2012.), but it is far from clear that they can.

This is a good place to remind readers that Shanahan (2016, and quoted above) said: '[The frame problem] in its technical [logic] guise, is *more-or-less solved*' (emphasis added). 'More or less solved' means not solved. So, even the logic version continues to bedevil AI researchers.

Don't all these points about logic and the so-called logic version of the frame problem mean that there's a lot more going on with the frame problem than merely getting the logic right (which in any case, we haven't got right)?

Is there even such a thing as getting the logic right? If so, why is getting the logic right so elusive? Could the elusiveness of getting the logic right be related to the fact that logic cannot capture relevance ('If President Obama is a prime number, then 2 + 3 = 19')? Could the difficulty of getting the logic right be related to the difficulty of finding an 'economical and principled way of succinctly saying what changes an action makes, without having to explicitly list all the things it doesn't change as well'?

An economical and principled way of succinctly saying what changes an action makes, without explicitly listing all the things it doesn't change would have to be a *metaphysical and epistemic principle of relevance*. This principle would make explicit what is relevant to what throughout our universe. The formal expression of such a theory would be a 'FORMAL EXPRESSION [of] our most favoured [and most strongly supported] inductive estimate of the world's taxonomic structure. Well, when we have [such a theory and such a formal expression], most of science will be finished' (Fodor, 1987, p. 147, emphasis in original).

So, the solution Hayes envisages seems to require the completion of science, at a minimum. Is the completion of science even possible? Is it even coherently conceivable? Science makes progress, but is it progressing toward explaining everything? Could humans even understand such an explanation? And even if science is completable, how would we know when we have completed it? And assume science is completable, how do we correctly update our metaphysics and epistemology? Isn't this a new version of the frame problem? Is there really a solution to the frame problem? Do humans actually solve it? (As mentioned previously, it is now known that humans (and indeed all thinking beings) do not solve the frame problem in any general way at all (see Dietrich and Fields, 2020). What Hayes wants cannot be got. Even with a 'completed science' (whatever that might mean), the frame problem is not solvable in any general way. So what Fodor wants also cannot be got. The best that thinking beings can do is to solve small, local, circumscribed versions of the frame problem. This we have all been doing since life arose on Earth. But no thinker solves this problem in general.)

Finally, isn't the frame problem dissolved once we take *empirical learning* and *empirical knowledge* seriously? Cognitive agents, from snails to humans, have learned about the ways of the world via *learning*. Isn't it this learning (and associated abstractions and generalizations) that keeps at bay worries about Antarctica's continued existence when you finish a cup of coffee? Isn't it this learning that keeps at bay questions about a pencil's colour given that you have just sharpened it? But doesn't such robust and deep learning require architectures inconsistent with the logic-friendly symbol processing architectures we discussed in Chapter 4?

Naturally, a fan of the frame problem would say here: 'Of course empirical learning blocks such irrelevant questioning, but the *real* frame problem is the problem of asking *Should empirical learning be the sole determiner of what's relevant to what?*

And so it goes.

These are all philosophical questions related to the frame problem. Some raise the issue of the ongoing failure to find a general solution to the frame problem. And some raise the issue that this failure points to something deeper going on. The problem of relevance is something deeper – a lot deeper (see Supplement 6.1). Still other questions are *limit questions,* questions about how far human understanding and human science can be pushed, questions about the limits of human knowledge. They show what searching for an economical and principled solution to the frame problem commits one to – wasting time. Finally, it is clear that McCarthy and Hayes were looking for a formal expression, in some logic, of a *metaphysical* and *epistemic* solution to the frame problem.

The frame problem, then, is a philosophy problem, a deep and undecidable problem discovered more than half a century ago by two AI researchers enamoured, reasonably enough at the time, with logic. The logic version of the frame problem is revealed to have been the tip of a large and dangerous iceberg: *the problem of relevance.* Having been found in a logic research program, AI researchers mistakenly thought the frame problem was a mere logic problem. But philosophers took up the frame problem and used it to explain why AI had thus far failed to build anything even remotely intelligent (because no AI algorithm correctly captures relevance as well as humans do). Naturally, this resulted in a war, with AI researchers saying that they were innocent of this 'philosophy frame problem', and with philosophers pressing their accusations. (There is even a well-known paper with the title 'Why AI is innocent of the Frame Problem'. See McDermott, 1987.)

The best example of a philosopher accusing AI of ignoring the gravity of the frame problem is this.

> God, according to Einstein, does not play dice with the world. Well, maybe; but He sure is into shell games. If you do not understand the logical geography of the frame problem, you will only succeed in pushing it around from one shell to the next, never locating it for long enough to have a chance of solving it. This is, so far as I can see, pretty much the history of the frame problem in AI; which is a main reason why AI work, when viewed as cognitive theory,

strikes one as so *thin*. The frame problem – to say it one last time – is just the problem of nondemonstrative inference; and the problem of non-demonstrative inference is – to all intents and purposes – the problem of how the cognitive mind works. I am sorry that McDermott [his 1987, see the paper title immediately above] is out of temper with philosophers; but, frankly, the frame problem is too important to leave it to the hackers.

Fodor, 1987, p. 148, emphasis in original

(Nondemonstrative inference is just coming to a conclusion based on evidence, rather than coming to a conclusion that must be true, given the premises, as in mathematics and logic. Nondemonstrative inference is extremely common. Scientists and police detectives and lawyers use it constantly. So do ordinary people – every hour of every day.)

For Fodor, the frame problem is the problem of how we draw inferences from data (evidence) and update what we know, accordingly. But this *is* McCarthy and Hayes's frame problem. Solving their logic-fluent problem is solving a version of Fodor's problem in miniature. Fodor states this when he says (to quote it again from above): 'The formal [i.e. logical] expression of such a [solution to the frame problem] would be a 'FORMAL EXPRESSION [of] our most favoured [and most strongly supported] inductive estimate of the world's taxonomic structure'. And Fodor gives an assessment of how deep this problem runs: '[W]hen we have [such a theory and such a formal, logical expression], most of science will be finished' (Fodor, 1987, p. 147, emphasis in original).

But matters are much worse for Hayes than merely being wrong that the *real* frame problem is the logic version. Recall from section 5, what McCarthy and Hayes were looking for: 'One *feels* that there should be some economical and principled way [e.g. an algorithm from logic] of succinctly saying what changes an action makes, without having to explicitly list all the things it doesn't change as well' (emphasis added).

This very *feeling* that Hayes is talking about is his assessment of what is *relevant* to the change in his research life of discovering, in 1969, the frame problem (in its logic guise). Hayes, in struggling with the frame problem, is doing exactly what Fodor says is the *real* frame problem: correctly inferring solutions from evidence and updating accordingly. Hayes is convinced that the solution to the frame problem lies within the set of *economical and principled* algorithms. Why is he so convinced? Because, for Hayes, only those types of algorithms are *relevant* to solving the problem. (Plus, of course, such an economical and principled algorithm would have major pragmatic benefits, like being easier to understand and render as computer code.) Quite interestingly, Hayes goes on to say that there doesn't seem to be such an economical and principled algorithm. Hayes, then, is basically admitting that *he's a victim of the frame problem*. He's looking in the areas he thinks are relevant and coming up empty, but he never changes what he thinks is relevant. (In section 8, below, we discuss other cases of falling victim to the frame problem.)

This completes our study of the two-version narrative of the frame problem: there is only one version, but it can be couched in many different ways. We now also know what the problem is, how it got its name, and how deep it runs. And we know that it is undecidable. It is obviously a philosophy problem.

Now we turn to two questions. (1) What are the real-world consequences of AI researchers ignoring the 'philosophy' version of the frame problem? (2) And what, if anything does this ignoring have to do with the fact that AI is populated only with monointelligences?

7. How Ignoring the 'Philosophy' Frame Problem Limits AI to Monointelligences

We can now bring two key ideas together: the *frame problem* and *monointelligence*. We have repeatedly used the term 'monointelligence' to refer to successful AI programs — programs that, e.g. play chess, heads-up no-limit Texas hold'em, or *Jeopardy!* — but which are not remotely as intelligent as a human (or a beetle). Even useful, money-making programs like the deep learning machines behind Facebook and Google are monointelligences. But humans and dogs and cats and birds and beetles are open-ended intelligences.

Monointelligences, by definition, do not have the frame problem. Not even AI machines like DeepStack, which can play imperfect information games like Texas hold'em, suffer from the frame problem. Why? Because they do not inhabit what we called in sections 7 and 8 of the Third War, *cognitive worlds*. As stated there, cognitive worlds are *expandable*, and do in fact expand – easily – growing to encompass many different types of tasks, all of which have to be dealt with using one type of mind: the kind that evolved here on Earth. Hence learning of an especially robust sort is required. Monointelligences, on the other hand, inhabit *task domains*. And task domains are *not* expandable. Task domains are highly circumscribed domains where only certain types of thoughts and actions are required.

What guarantees the circumscription is the task. If the monointelligence makes diagnoses of heart disease, then that's the task. If it plays Texas hold'em, then that's the task. And it is the *only* task. It is therefore relatively easy to guarantee that changes occurring in the task domain result in correct updating of the appropriate and relevant beliefs and other items in memory. The relevant database might be quite large, but it is still very small compared to what an ordinary human carries around in her memory. So, the database for the task domain can be effectively partitioned or, if need be, all of it can be examined. Hence, in task domains, there is no frame problem. Basically, monointelligences do not lead complicated enough existences to have the frame problem. It should be noted, though, that nowadays, most modern monointelligences learn. This is because (1) it makes for a better working machine, and (2) it is easier to program. But again, the learning is not vastly multidimensional like the learning required of animals.

Therefore, AI's great successes are due to the central fact that AI programs do not have the frame problem. The AI programs do only one thing. The machines do their one thing well, and that's why no one can beat AlphaZero at Go. But it is *one thing*. And only *one thing*. When the game is over, AlphaZero cannot drive home and feed the children and cats and dogs.

8. The Dark Truth about the Frame Problem: There is no Solution, but there are Victims

Both early AI researchers and philosophers sought a solution to the frame problem – a way to sideline the problem for good. Our reader can see this in the quotes from McCarthy and Hayes and Fodor. Of special interest, however, is how humans solve it: How do we humans algorithmically and efficiently determine what information is *relevant* to a given change?

Well, how do humans solve it?

Answer: *We don't because we can't.* We guess and hope for the best. So, the search for a 'solution' is itself a waste of time.

To make our case that we humans, when experiencing a change, don't update just those beliefs and items of memory that need updating, it suffices for us to note that humans and other living things are constantly *victims* of the frame problem. Here are some examples. We start with the shocking, huge ones, and then consider a couple more mundane ones. In all of these examples, humans didn't (and don't) update their beliefs and other memories that they should have – humans are victims of the frame problem. At the time, the beliefs or understandings, etc. that they didn't update seemed irrelevant, and so justifiably ignored. But such ignoring proved catastrophic, when it wasn't merely annoying.

8.1. Global Warming

Global warming started to accelerate with the industrial revolution (approximately, 1760 to 1830). From then through to today, global temperatures have climbed. Earth is heating up, polar ice caps are melting, and the weather is changing. Global warming is a stunning example of the frame problem on a world-wide scale. No one updated their beliefs and understandings about what life would be like if billions of cars were driven around everyday, while factories of all kinds pumped CO_2 into the air and while rain forests were razed around the globe. True, in the 1960s, some concerned scientists finally warned the world that global warming was a likely result of industrialization, but no one in government or industry listened (intentional victims of the frame problem?), and anyway, by then it was too late.

8.2. Global Warming Denial

A related large example is that few predicted the rampant denial of global warming by whole governments and large majorities of citizens. Even though we all knew that a majority of Americans deny that evolution happened and that vaccines are good for one's health, still, few foresaw that global warming denial would be the backbone of a major political party.

8.3. Darwin and Evolution Denial

Darwin foresaw and even predicted the negative effect his theory of evolution would have on Christian England in the nineteenth century. This is why he waited twenty years to publish his theory, and then he published only because he was about to be scooped by Alfred Russell Wallace. But even Darwin didn't foresee the robust resistance to his explanation we see today around the world and in countries that once strongly embraced science. Darwin did update his beliefs about the effect of evolution, but only for his generation and his country.

8.4. Antibiotics

Even though most physicians and drug researchers are taught and understand evolutionary theory, few foresaw the evolution of robust pathogens due to overuse of antibiotics. Yet this is what has happened.

8.5. Dangerous behaviour

It is perhaps reasonable to include other medical examples. Tobacco use is one case – especially smoking it. As people generally got healthier, the detrimental effects of tobacco use became more obvious. But again, few updated their beliefs about tobacco, recognizing that it was as deadly and addictive as it has turned out to be. Not using seat belts while driving is another case. Driving while talking or texting on one's cell phone is another. In all these cases, a serious public health threat emerged that was unforeseen by those who should have foreseen it.

8.6. Mundane Examples

Have you, dear reader, ever lost your keys? Most cases of losing one's keys are examples of being victimized by the frame problem (putting aside unknown holes in pockets and the like). You put your keys down, but not in a spot you normally put them. You don't update your beliefs correctly, and *Bingo!* – lost keys. Losing wallets and cell phones are also frame problem examples.

People back out of their garages without raising the garage door. They start fires in fireplaces without opening the flues. They drive away from fuel pumps with the pumps' nozzles still in their cars' fuel intake pipes. And so on and so forth.

Being a victim of the frame problem is just part of living. If your mistake is bad enough, the consequences will be too. But usually we find our keys. The fact that we are often victims of the frame problem strongly makes the case that, given an experienced change, what humans and other animals do is update what *seems to them* reasonable candidates for updating, while ignoring the rest. However, ignoring candidates for updating often excludes candidates that should have been updated. Which is just what you would expect if humans' approach to the problem were *guess and hope for the best*.

Such a 'solution' to the frame problem is called a *heuristic* solution – a solution that relies on heuristics. A heuristic for solving a problem is a trial-and-error method that is not guaranteed to work. It works some, maybe most, of the time. But it doesn't work all the time. Heuristics are used all the time by all animals with even moderately complex brains inhabiting even moderately complex cognitive worlds (see the Third War, sections 7 and 8).

Fields (2013) shows that re-identifying any object, including any other person, as the same object encountered at some previous time requires solving the frame problem. People often misidentify objects, mistaking new ones for old ones or vice versa. All such errors are due to the frame problem. So, the problem is not just widespread, it is ubiquitous.

9. The Bright Truth about Humans: We have ways of handling the Frame Problem . . . sometimes

Finding a solution to the frame problem, i.e. never falling victim to it, would require knowing how everything in the universe is connected. That's impossible. Not only is such information far too large for our brains (huge though they are), but each event is new and establishes *new* connections, and we are mostly ignorant of these new connections. So, we can't solve the problem, rather, as we said above, we guess and hope for the best. But we are not guessing blindly or by flipping a coin. Our guesses often are informed, intelligent. We can be pretty good at guessing. How?

On the current leading theory in cognitive science, humans think about things by deploying *concepts* representing those things (Margolis and Laurence, 1999; Murphy, 2002; and Margolis and Laurence, 2014). And each individual human's concepts are all connected in one vast network. This network embodies each human's panoply of what is relevant to what. The more distant the connection between two concepts in a person's brain (say, between penguins and the Pythagorean Theorem), the more irrelevant (probably) the two are to one another. The 'probably' is just the recognition that humans calculate relevance using probabilities – the probability that penguins and the Pythagorean Theorem are relevant to each other is low, so they are quite distant from each other in the concept net. But, as with all probability measures, conceptual relevance measures are defeasible – constantly open to revision. And revise we do, as we time and time again extricate ourselves from some new case of falling victim to the frame problem. Such revision is a kind of *learning*. We learn that increased carbon dioxide in the atmosphere is due to driving cars and running factories and felling forests, we learn that lids on boiling pots of water also get hot, we learn that we should always put our keys on the peg by the door to avoid having to tear the house apart looking for them and so forth (there may be more than just learning going on. See Supplement 6.1).

This is a beautiful picture and a beautiful theory. But alas, this theory of concepts is mostly wishful thinking, for the general theory of concepts is in complete disarray. The shocking truth is that, though they are central to most theorizing in cognitive science, no one knows what concepts really are (see the citations in the paragraph above, and see Machery, 2009, 2010). This is similar to the theory of genes before the discovery that genes were strands of DNA (usually in its double helix form). We used the notion of genes before we knew what they were. (Genes were first proposed as discrete heritable units coding for molecules that have specific functions in living things by Gregor Mendel from research he did this during the years 1857 to 1864. However, Mendel didn't use the term 'gene'. That term wasn't introduced until 1905. And though DNA, the molecule, was discovered in 1869, it wasn't until the 1940s and 50s that researchers figured out that genes were strands of DNA. Finally, it wasn't until 1953 that the structure of DNA – a double helix – was discovered.)

Nevertheless, we can be very confident that one day we will develop a theory of how our brains and minds work, and so how we manage to dodge the frame problem from time to time by updating what's relevant in our memories. If AI and neuroscience are any guide, it is very unlikely that this eventual theory will be even a distant cousin of the theory of networked concepts. But who knows? What is exciting to contemplate is that once we have such a theory, we might be able to use it to tell us how to augment our brains and program our machines so that we all fall victim to the frame problem far less that we do now. That would be a stunning advance. The future is bright . . . as long as we update our relevant beliefs.

Fourth War Supplement 6.1

Analogy, Creativity, and the Frame Problem

We have argued that the Frame Problem is really the problem of determining what information is *relevant* to what. This is a philosophy problem because, like all philosophy problems, it runs deep; running deep is the telltale mark of a philosophy problem. Making progress on the problem of relevance (as we will call it) requires (1) advances in the problem of meaning (semantics), for this is how relevance is determined, and (2) advances in understanding how humans think (for this is how relevance determination is implemented). The latter of these two is a science problem – specifically, a problem for neuroscience and psychology. Though this problem is one of the deepest in neuroscience and psychology, it nevertheless remains a science problem. But the former problem, the problem of meaning is a central philosophy problem (see Chapter 5 on the Third War). Why? Because this problem requires understanding consciousness, since meaning and semantics are bound up with it. So, relevance is bound up with consciousness. And consciousness appears to be the deepest problem of all (see the opening lines of Chalmers, 1996).

However, the frame problem is richer still. Humans routinely and seemingly spontaneously find new connections between their concepts (in spite of section 9, we are going to use the term 'concepts', mainly because there is no other term). There's a special process involved; this process is called *analogy-making*. In making an analogy, a human finds a connection between two concepts that were previously unrelated. More specifically, by *analogy-making*, we mean a human's ability to see something, X, as like a Y, where X and Y can be, and usually are, quite semantically distant from one another.

Analogy-making is quite common among humans, as well as spontaneous. It is also a central example of *creativity* – analogy has played a key role in science. For example, Darwin's insight about how evolution used natural selection was in part fuelled by his understanding of how humans had engaged in artificial selection of animals and plants.

Consider the following typical (and real) example: Two people are cross-country skiing together. They pause to rest and drink some water. Though it is cold, they are quite warm and, being mindful of hypothermia caused by overheating, one of the pair takes off one glove to cool down. She then quips: 'My hand is like a dog's tongue when he's panting in the summer.'

Fodor was quite impressed by this human ability and its relation to the frame problem. He said:

> Mainstream cognitive science has managed to get the architecture of the mind *almost exactly backwards* [emphasis in original] ... By attempting to understand thinking in terms of a baroque proliferation of [AI programs using] scripts, plans, frames, schemata, special-purpose heuristics, expert systems, and other species of domain-specific intellectual automatisms – jumped-up habits, to put it in a nutshell – it missed what is most characteristic,

and most puzzling, about the higher cognitive mind: its ... creativity, its holism, and its passion for the analogical.

1985, p. 4

The almost-exactly-backward comment is even more contentious than it appears, because many researchers consider analogy research to be one of cognitive science's greatest success stories: there is a robust theory of analogy. Furthermore, theory-based computer models of analogy-making offer some plausible explanations of this central human cognitive ability.[8] Still, Fodor's main message in the quote is well-taken: The human mind does have a passion for the analogical.

However, somewhat surprisingly, Fodor used the existence of analogy-making as an anti-AI argument, arguing that this passion for the analogical cannot be implemented on a computer. 'Somewhat surprisingly' because Fodor has endorsed the computational theory of mind:

> [The computational theory of mind] is, in my view, by far the best theory of cognition that we've got; indeed, the only one we've got that's worth the bother of a serious discussion ... [I]ts central idea – that intentional [semantical/meaning-based] processes are syntactic operations defined on mental representations – is strikingly elegant. There is, in short, every reason to suppose that Computational Theory is part of the truth about cognition.

Fodor, 2000

The rub, however, is that, whereas quite a few AI researchers and cognitive scientists think that computationalism is either *all* or *most* of the truth about how our minds work, Fodor thinks it is only a small part of the truth. Among the parts of the truth left out are the parts about analogy (Dietrich, 2001b).

What is it about analogy that makes it non-computable on Fodor's theory? First, let's briefly examine Fodor's theory of mind. Then we will, again, briefly, examine analogy.

Fodor's Modularity Theory of the Mind

Fodor's theory of the mind is that it comprises two very different kinds of information processors: modules responsible for perception (primarily vision, hearing, tasting, feeling, smelling) and for some aspects of language (e.g. parsing sentences to figure out the subject and the predicate), and a non-modular central processor (Fodor, 1983). All modules have evolutionarily narrow, fixed jobs and relatively fixed neural architectures (and most have more or less specific locations in the brain). Modules are all domain-specific (e.g. figuring out shape or texture of a perceived item), they are relatively fast, they process only one kind of information, and, most importantly, they are all informationally encapsulated. A module can be construed as a 'special-purpose computer [like a calculator] with a proprietary database [where], (a) the operations that [the module] performs have access only to the information in its database (together, of course, with specifications of currently impinging proximal stimulations); and (b) at least some information that is available to at least some cognitive process is not available to the module' (Fodor, 1985, p. 3; his emphases). The notion of information encapsulation is captured in points (a) and (b), above.

The persistence of visual illusions such as the famous Müller-Lyer illusion provide hard evidence that vision is a module.

The two lines are the exactly same length. Measure them (or copy them and paste one on top of the other). The bottom is *not* longer than the top one. And the top one is not shorter. Yet, armed with this information (which resides in your central processor), your vision system persists in producing the illusion: you cannot *not* see it. The information that the lines are equal length does not block seeing the illusion. Hence, your vision system does not have access to clearly relevant information – if it did, you'd see the lines as the same length. Hence your vision system is a module. The information is there, but it is unavailable to your vision module.

The central processor, on the other hand, is *not* informationally encapsulated. This is its defining property, according to Fodor. The central processor's main job is producing evidentially supported beliefs. So, the central processor is slow, has access to everything the person knows and believes, and is not domain-specific.

More technically, the central processor, according to Fodor, is rational in an ideal sense: it is both *isotropic* and *Quinean*. Producing evidentially supported beliefs is isotropic if and only if a non-arbitrary sample of all the

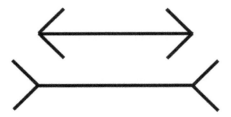

Figure 6.1 The Müller-Lyer illusion. The two horizontal lines are the same length, though the top one appears to be shorter.

relevant evidence is considered. Producing beliefs is Quinean if and only if the level of confirmation of any one belief is sensitive to the level of confirmation of any other belief and to all the beliefs taken collectively. As Fodor puts it, when it comes to producing beliefs, 'our botany constrains our astronomy' (Fodor, 1983, p. 105).

Now for some conclusions. Modules do not suffer from and cannot be victims of the frame problem because they are informationally encapsulated – *modules are monointelligences*. But the central processor can and does. Also, analogy-making is something only one's central processor can do. In fact, analogy-making is part of what makes the central processor isotropic and Quinean. Finally, nothing that is isotropic and Quinean can be a computer of any sort (Dietrich and Fields, 1996). So, Fodor is committed to a kind of anti-AI position: the central processor cannot be implemented as an AI, a position he seems to embrace (Fodor, 2000; see Dietrich, 2001b for a rebuttal).

Analogy-making is truly marvellous, but it is not magical. On the available evidence, analogy-making is computational. So, what is analogy-making? Here is a very brief description.

Analogy-making

Many of the analogy researchers (those cited in endnote 8 and many others) talk as if the freshness of an experience of analogy resides solely in seeing that something is like something else – seeing that the atom is like a solar system, that heat is like flowing water, that paint brushes work like pumps, or that electricity is like a teeming crowd. But analogy is more than this. Analogy isn't just seeing that the atom is like a solar system; rather, it is seeing something *new* about the atom, an observation enabled by 'looking' at atoms from the perspective of one's understanding of solar systems. The questions analogy researchers should be asking are: (1) Where does this new knowledge about atoms come from? (2) How can an analogy provide new knowledge and new understanding?

One answer is that having an analogy *changes* the concepts involved in the analogy – merely having an analogy changes one's concepts (Dietrich, 2000).

One way such change could happen is this. Suppose that a process (we will call it A, for analogy) was always available that applied concepts at various psychological distances from concepts of immediate interest to those very concepts of immediate interest. The job of A is to produce abstract similarities between the concepts (psychological distance is just *semantic distance*: the concepts for ice cream and cold are closer together than those for ice cream and the formula for the area of a triangle). Process A focuses on concepts of immediate interest, say a case of trying to solve a problem or figure something out, but A could also just run in the background. Furthermore, suppose that when two concepts, $C1$ and $C2$, interact in a mind, the product of their interaction is an *abstract* version of both, with the immediate concept dominating the abstraction interaction – that is the amount of change is not symmetric between $C1$ and $C2$. So, if $C1$ is the immediate concept of interest and $C2$ is the distant concept, then when A brings $C1$ and $C2$ together, $C2$ is abstracted in such a way to more closely resemble $C1$. And lastly, suppose that the greater the psychological (semantic) distance between $C1$ and $C2$, the more abstract the product of $C1$ and $C2$ is. Then when solving a problem, A is going to produce new ideas and new concepts (new versions of $C2$) – which is a kind of creativity. And these will be available for updating the central processor's database of beliefs and knowledge, which updating, of course, will be partial.

There is good psychological evidence for process A. Also, more concrete ideas for how A would work have been developed (see Dietrich, 2010). It seems clear that A helps in a specific way with the frame problem: by producing new, abstract concepts and adding them to a person's belief database, A is producing new ideas that are useful for dealing with current and future problems, as well as producing wider understanding for the relevant person. But it is also clear that A doesn't *solve* the frame problem. In fact, misplacing our keys is the sort of problem analogy-making *cannot* solve. So, no analogy-making system is going to solve the frame problem in its full generality. In fact, nothing is. But processes like A give us creativity and insight. So, though we will misplace our keys from time to time (and worse, unfortunately), we will also produce great science.

Part II

Beyond the AI Wars:
Issues for Today

Introduction

Wars require passion. The AI wars would not have been fought had the participants on both sides not felt strongly about it. Plenty of fuel was available to ignite passion, from the thrill of discovery and promise of entrepreneurial success on the pro-AI side to worries about job displacement or dystopian forecasts of robot takeovers – the HAL 9000 in Kubrick's *2001, A Space Odyssey* appeared in 1968 – on the anti-AI side. The main fuel for the fire, however, was something far deeper. Understanding it requires going back 500 years, to Copernicus and the dawn of the Scientific Revolution.

Though the idea that the visible planets orbit the Sun was known in ancient Greece, it was forgotten until the Renaissance. Copernicus reintroduced it at the right time and place to trigger a cultural revolution. Northern Europe was convulsed by the Reformation when *De Revolutionibus Orbium Coelestium* appeared in 1543, though the brutal century of religious warfare had barely started. Europe in the 1540s was still stunned by the New World, Magellan's circumnavigation of the Earth, and the meteoric rise in Spanish wealth. Traditional truths were everywhere being overturned. Traditional structures of authority and power had lost their balance and would soon topple. Displacing the Earth and its inhabitants – displacing *us* – from the centre of the universe fit the sixteenth century zeitgeist. The economic displacement of the Industrial Revolution followed by a mere two centuries, and Darwin's biological displacement by only a century after that.

It is not only the traditionally religious that value the idea that humans are special, not just cognitively but cosmologically. As cultural historian Yuval Harari (2014) has emphasized, human exceptionalism has been the foundational cultural myth from the Agricultural Revolution (ca. 10,000 BCE) onward. For partisans of this myth, the narrowing of possibilities for human specialness instigated by the Scientific Revolution was and has continued to be intensely painful. For the iconoclasts at the forefront of scientific and technological disruption, each assault on human specialness has been an intellectual – and often an economic – triumph.

AI presents an existential threat to human specialness. Like *De Revolutionibus Orbium Coelestium*, AI arrived at an auspicious cultural moment. World War II had confirmed the human penchant for violence and evil in ways that could not be ignored. Nuclear weapons demonstrated to all that humans could, if provoked, commit planetary suicide. The sciences, however, had entered a period of extraordinary optimism, bolstered in part by the post-war creation of institutionalized research-funding mechanisms.

The transistor (1947), the structure of DNA (1953), and the Salk vaccine (1955) are among AI's contemporaries. The 'cognitive revolution' and the decline of Behaviourism in psychology can be dated to Chomsky's scathing review of Skinner's *Verbal Behavior* in 1959. AI took the re-acknowledged mental processes, the most obvious evidence of human specialness, and proceeded to automate them. Lucas' (1961) argument from the 'infinite variety of the human mind' to not just the failure but the pointlessness of AI expresses the horror felt by defenders of specialness.

From this 500-year perspective, the AI wars of the later twentieth century can be seen as a desperate rearguard action in a much larger conflict. The turn of the twenty-first century, moreover, brought a new player on the scene: human neuroscience grew from an obscure medical specialty to front-page news with the introduction of neuroimaging. People could suddenly, literally, and bafflingly *see* their minds in operation. Something previously completely mysterious was revealed, and scientists talked excitedly about reading the mind by watching a movie of the brain's activity.

The rise of neuroscience had a profound effect on the broader discipline of cognitive science. During the period of the AI Wars, cognitive and developmental psychology made impressive gains in experimental technique and generated far more data about both the adult functioning and the development from infancy of human cognitive and emotional capabilities. For theories of cognition, however, cognitive science turned to AI. Symbolic and neural network-based theories appealed to different research communities within psychology, many members of which were active participants in the Second War (see Chapter 4). Neuroscience changed this. Neuroscience provided psychologists with a direct means to visualize what the brain was doing during standard experimental tasks. Neuroscience also provided names for the functional 'modules' of cognitive psychology: the names of brain regions, and later the names of functional networks that were active during some tasks and not others. Psychologists no longer needed the abstract vocabularies of symbol processing or ANNs: they now had an anatomical and functional language for the brain itself. Cognitive neuroscience was born.

Not only the rise of cognitive neuroscience, but also successes within AI itself, helped to create a new landscape for philosophical discussions about AI in the early 2000s. Machine learning made rapid advances, providing new capabilities to analyse 'big data' in feasible timeframes. Small-scale AI applications became ubiquitous, even in popular culture. People started talking to their phones, and now many talk to their houses. Robots took over manufacturing, and even started driving cars.

We examine two of the main effects of these changes here in Part II: the re-invigoration of research on consciousness, and the rise to prominence of ethical concerns about AI. They raise a question hardly considered during the AI wars: Is human-like intelligence really the goal of AI? What would human-like AIs be *for*?

7

What about Consciousness?

---

1. Introduction

The most fiercely defended redoubt in the battle to defend human specialness is the very core of human mentality: consciousness and its surrogates, meaningfulness and aboutness. As discussed in Chapter 5, consciousness appeared in the Third War mainly in the guise of aboutness. The debate was, in that setting, mainly about *semantics*: how could symbols relate to things? Neuroscience moved the focus of this debate from aboutness to awareness. Part of the motivation for this refocusing was practical: neuroscience has deep roots in medicine, and a crucial medical question, particularly in surgery, is whether the patient is aware of anything at all.

The new focus on awareness and the new language of neuroscience introduced new questions to the debate, some of which could hardly be formulated in the earlier language. While most philosophers are well-versed in logic and some have backgrounds in computer science, hardly any have training in biology, much less the arcana of current neuroscience. Biologists, for their part, are no more likely than computer scientists to have any more than superficial knowledge of philosophy. Hence the perennial problem of finding acceptable terms of debate has become even more severe in the twenty-first century than it was in the twentieth.

The 'consciousness redoubt' has not yet fallen, despite the attack being reinforced by legions of neuroscientists in the new millennium, and despite the re-emergence of an ancient and long-neglected threat to human specialness: the panpsychist possibility that consciousness is everywhere. This chapter asks why, and whether it matters.

2. Intuitions and Zombies

The Third War introduced the primary philosophical argument in the battle for consciousness: the 'zombie argument' developed by Chalmers and others (see especially Supplement 5.4). The argument, recall, goes like this: Faced with a claim that X explains consciousness, imagine a creature, computer, or whatever seems appropriate, that has X but is unconscious. Voilà: X isn't *logically sufficient* for consciousness, so X can't explain consciousness.

It is worth understanding this argument in detail. To do this, it helps to formalize the argument as a schema. Let **P** represent some claim, let **M** and **L** represent, respectively, the ***Possibly*** and ***Necessarily*** operators of standard modal logic, and let '¬' represent negation[1]. Then the *logical possibility argument* schema is the following:

Claim: **P**
Response:
Premise: 1) $M\neg P$
Hence: 2) $\neg LP$
Premise: 3) $\neg LP \to \neg P$
                   4) $\neg P$

The premises of this response are 1) and 3). Premise 1) claims that it is *logically possible* that **P** is false, from which it follows that it is not *logically necessary* that **P** is true (step 2). Premise 3) claims that unless **P** is logically necessary, it is not true. Hence by *modus ponens*, ¬**P**.[2]

Premise 1) claims a logical possibility: $M\neg P$. Showing that something is logically possible is easy: if it can be *coherently* imagined, it is logically possible. For any proposition **Q**, $M\mathbf{Q}$ is true unless **Q** is a logical contradiction. Here are some examples:

i) Let **P** = 'A computer implementing program *C* understands Chinese'. A 'computer' comprising a person implementing *C* by hand in a room, however, can be coherently imagined and it seems obvious in this case that neither the person nor the room understands Chinese, so it is logically possible for a computer implementing *C* not to understand Chinese. Note what is going on here: we can *imagine* a person in a room implementing *C* and the person *seems to us* not to understand Chinese. Why does the person implementing *C* seem not to understand Chinese? We *imagine* that *we* are that person in the room implementing *C*, and this imagined activity does not *feel like* understanding to us. If we assume that understanding has to *feel like* understanding, it follows that our imagined selves do not understand Chinese. Hence *M¬P* is true: it is logically possible that a computer implementing *C* does not understand Chinese. Therefore, following the logical possibility argument schema above, ¬**P**. This is Searle's (1980) Chinese Room argument against AI language understanding (see Chapter 5 on the Third War).

ii) Let **P** = 'A robot pencil sharpener is conscious'. However, it is at least imaginable that a robot pencil sharpener is not conscious; this was coherently imagined, albeit briefly, in The Third War (Chapter 5). Hence, following the schema, ¬**P**. This is a standard zombie argument against AI consciousness: unless something *has to be* conscious, as a matter of logic, it is not conscious. Most people do not, for example, believe that robot vacuum cleaners or self-driving cars are conscious. If we do not *have to* assume they are conscious to explain their behaviour, we don't. While some, most notably Hameroff and Penrose (2014), have suggested that quantum computing might implement consciousness where classical computing fails, zombie arguments equally apply to this case, too: we don't *have to* assume that quantum computers are conscious to explain their behaviour, so by the logical possibility argument, they are not conscious.

iii) Let **P** = 'If I am awake and my nervous system is functioning properly, I am conscious'. We can, however, imagine my zombie twin, who replicates my nervous system and its functions perfectly but is not conscious. Therefore, following the schema, ¬**P**. This is Chalmers' (1996) zombie twin argument against any neuroscience-based explanation of consciousness (see Supplement 5.4 to The Third War). It directly challenges the assumption that we *have to* assume other humans are conscious to explain their behaviour. Hence it demands that we come up with some other, non-behavioural or more generally, non-functional reason for supposing that our fellow humans are really humans, not zombies.

iv) Let **P** = '*a* displaying evidence X implies *a* is conscious' for some entity *a*. Here X can be any evidence at all, detected by any public means; X could, for example, be a particular pattern of brain activity detected by EEG or fMRI, or the ability to drive a car at high speed through heavy traffic, write poetry, display empathy, discover new theorems, or cook a good meal. If, however, we can imagine *a* displaying behavioural evidence X without being conscious, or some creature or entity sufficiently like *a* displaying behavioural evidence X without being conscious, we can again employ the schema and conclude ¬**P**. This is a general argument against any empirical evidence whatsoever being evidence for consciousness (see Dietrich and Hardcastle, 2004), again as discussed in The Third War. It is, moreover, a general argument against any kind of 'Turing Test' for consciousness, no matter how sophisticated (see Chapter 2).

The second premise (step 3) of the logical possibility argument schema, ¬**LP** → ¬**P**, asserts that **P** has a particular property: it can only be true if it is necessarily true, i.e. if it cannot, as a matter of logic, be false. In every case the proposition **P** asserts that some entity *a* has some property *F* (this is written **P** = *Fa*), so this second premise is effectively the claim that this *a* must necessarily have *F*: *a* would not be *a* without *F*. Properties of this sort are *essential* or *defining* properties of *a*, properties that one would use to identify *a* as *a* and not something else (for accessible reviews of a huge literature, see Scholl, 2007; Fields, 2012). Kripke (1972) pointed out that in some cases, essential properties can be discovered empirically; for example, one can discover that water is H_2O, after which being H_2O, and not something else, is an essential property of water. For this to happen, it must be possible to independently, using different observational methods, determine that something is water and determine that it is H_2O. One must then be able to develop a theory that explains why H_2O is water and why water is H_2O. The claim that water is H_2O, essentially, is then the claim that this theory is true.

The trouble with consciousness is that we lack generally acknowledged, public, observational methods for determining whether an entity is conscious. If we had such methods, there would be no question about whether robot pencil sharpeners or self-driving cars were conscious; we could find out by looking. Until surprisingly recently, it was a common belief, among academics at least, that nonhuman animals and even human infants were unconscious; see Panksepp (2005), Rochat (2003), or Trevarthen (2010) for arguments, addressed in each case to academic colleagues, that animals and infants are in fact conscious. This previously widespread belief in the unconsciousness of animals and infants has substantially disappeared over the past decade or so. It is instructive, however, that it did not disappear because the consciousness of animals and infants was experimentally demonstrated. It disappeared in part due to ethical concerns, in part due to evidence from comparative functional neuroscience, and in part due to demonstrations of intelligent behaviour (cephalopods provide an interesting example of this latter case; see Godfrey-Smith, 2016). The kinds of behaviour typically indicative of consciousness in animals – learning, memory, attraction and aversion – are also observed in plants (e.g. Gagliano, Renton, Depczynski, and Mancuso, 2014) and even in micro-organisms (e.g. Baluška and Levin, 2016). Are these organisms conscious? Without a generally acknowledged observational method for determining whether an entity is conscious, any answer remains an indirect inference. As we concluded in Chapter 5, who knows?

In the absence of some means to empirically demonstrate that F is an essential property of a, the only way to make $L(Fa)$ true is to *assume* or *stipulate* that F is an essential property of a. In this case, having F becomes part of the *definition* of a. This is what has happened, *de facto*, with consciousness and humans: humans, at least non-infant humans, are regarded as *conscious by definition*. Why? Because this *feels* right. We are conscious, and it *seems to us* that other people are conscious, so they *must be* conscious. This protects us from the 'other minds argument' that the people surrounding us may be unconscious zombies. We do not have to discover *evidence* of their consciousness if they are conscious by definition. We can choose to assert that there *could be* no such evidence as long as we assume they are conscious by definition.

We see then, that if consciousness is a defining property of any entity or being a that is or could be conscious, the premise, $\neg L(Fa) \rightarrow \neg Fa$, must be granted when F is or entails 'is conscious'. Hence the underlying assumption of the logical possibility argument schema as it is applied to consciousness is:

Assumption: If **a** is conscious, consciousness is a defining property of **a**.

With this assumption, $Fa \rightarrow L(Fa)$ when F = 'is conscious' and a is any entity. Rendered in English, this is 'If a is consciousness, then it is necessary that *a* is conscious' (because consciousness is a defining property of *a*). Hence $\neg L(Fa) \rightarrow \neg Fa$ (this move is established by the well-known logical transformation: '(If A then B) is equivalent to (If Not-B then Not-A)').

This same assumption is generally made regarding consciousness-associated properties like understanding. People understand because understanding is a defining property of personhood. Hence the zombie arguments i) – iv) above can be restated, considerably less formally, by the following:

i') Running some program *C* is not part of the *definition* of language understanding, so a computer running *C* does not thereby understand language.

ii') A robot pencil sharpener is not *defined* as being conscious, so it is not conscious.

iii') Consciousness is not *defined* by proper neurocognitive functioning, so proper neurocognitive functioning is not consciousness.

iv') Nothing externally observable has been *defined* as evidence of consciousness, so nothing externally observable is evidence of consciousness.

This way of stating these arguments removes the imaginative intuition pumps of workers locked in rooms and agile but unconscious zombies, and displays them for what they are: arguments about how words like 'understanding' and 'consciousness' are *defined*.

If, as discussed in The Third War, words like 'understanding' and 'consciousness' have *only* first-person ostensive definitions, arguments i') – iv') appear plausible: *my* understanding and *my* consciousness do not appear to involve programs, particular neurocognitive functioning, or behavioural evidence. Science, however, is precisely in the business of replacing ostensive definitions, whether first- or third-person, with more tractable ones; hence 'water = H_2O'. Non-ostensive definitions can be either theoretical or operational. Here are two examples:

> Theoretical: 'Integrated Information Theory' (IIT; Tononi, 2008; Tononi and Koch, 2015) *defines* consciousness as integrated information, a precisely specified information-theoretic measure of the level of causal interaction within a network of processing elements. Any system with non-zero integrated information is, in the context of IIT, conscious *by definition*.

Operational: Anesthesiologists accept as sufficient evidence of, i.e. operationally define, unconsciousness during surgery as an absence of any patient-reportable during-surgery experience together with an absence of various behavioural and physiological signs (e.g. Bischoff and Rundshagen, 2011; Boly, Sanders, Mashour and Laureys, 2013). If a patient does not report during-surgery experiences or display any of the relevant behavioural and physiological signs, then they were unconscious during the surgery *by definition*.

If definitions such as these are accepted, the logical possibility argument schema clearly fails. If I have *defined* consciousness in terms of integrated information, for example, then I cannot coherently imagine a system having sufficient integrated information but no consciousness. I may *think* I have imagined such a thing, but my imagination conflicts with my definition and is therefore incoherent. If I have *defined* consciousness as the ability to report experiences, then I similarly cannot coherently imagine unconscious zombies that report experiences.

The zombie argument, the Chinese Room argument, and all other arguments that employ the logical possibility argument schema are, therefore, fundamentally arguments about how words are to be defined. They are, in particular, arguments from common intuition that non-ostensive definitions of words such as 'understanding' and 'consciousness' are philosophically or morally suspicious and hence not to be accepted. Human history amply shows that how words are defined can arouse passions and instigate warfare. The AI Wars are no exception.

3. Navigating the Linguistic Minefield

Do we really care whether self-driving cars can consciously see? Is it worth fighting a war over, even just a philosophical war? It depends, one might say, on what we mean by 'see'. Claiming that a self-driving car can see is not the same as claiming that it has human-like consciousness, but what is it claiming? Making progress on this question requires working our way through a minefield of hidden assumptions, ambiguity, equivocation, and incoherent or even contradictory usages. Fortunately, the explosive development of cognitive neuroscience in the twenty-first century allows considerably more conceptual clarity, backed in many cases by dozens to hundreds of experimental or clinical studies, than was available during the AI Wars. In many cases, this clarity comes in the form of experiments or clinical studies that demonstrate the preservation of some kinds of conscious experiences in the absence of others, e.g. the preservation of perceptual experience but loss of the ability to record such experiences in memory, characteristic of Korsakoff syndrome (Fama, Pitel and Sullivan, 2012).

Typologies of consciousness or of kinds of experiences are as old as the world's religions. For the present purposes, however, a good place to start is Damasio's (1999) distinction between protoself, core consciousness, and extended consciousness. Protoself, for Damasio, is an *unconscious* awareness of bodily state, the kind of awareness needed to maintain homeostasis and hence to survive. Core consciousness involves conscious awareness of bodily state, e.g. conscious awareness of pain or hunger. Extended consciousness is imaginative and involves the ability of 'mental time travel' (Suddendorf and Corballis, 2007) either backwards into episodic memories or forwards into planning for the future. The notion of *unconscious awareness* here should give us pause: its status as a prima facie oxymoron is a red flag. To make sense of this, it is useful to combine Damasio's typology with Block's (1995) distinction between 'phenomenal' and 'access' consciousness. Phenomenal consciousness of X is experience of X. Access consciousness of X is an ability of X to play a functional role in behaviour. Hence we have Table 7.1.

Panksepp (2005) further divides core consciousness into immediate sensory experience, whether of bodily state or the outside world ('primary-process consciousness'), and the interpretation of that experience, e.g. the inference of correlations between external and internal experiences ('secondary-process consciousness'). Primary-process without secondary-process consciousness enables one to experience putting one's hand in the fire and the pain of being burned, but not to consciously connect the two.

It is worth noting that the use of the word 'consciousness' in these definitions is already controversial. Many reserve 'consciousness' for

Table 7.1 Damasio's distinctions crossed with Block's, yielding a 3x2 typology of consciousness

Type of Consciousness:	Protoself	Core Consciousness	Extended Consciousness
Phenomenal	None	Bodily state, e.g. pain	Memory, plans
Access	Homeostasis	Real-time behaviour	Reflection, planning

human-like consciousness, considered as indivisible or un-analysable. Lucas (1961), for example, considered 'consciousness' to entail unlimited meta-consciousness: knowing that one knows that one knows . . . indefinitely. In a critique of IIT, Cerullo (2015) argues that Tononi equivocates between a use of 'consciousness' to mean a kind of bare experience that Cerullo calls 'protoconsciousness' and a use of 'consciousness' to mean the kind of behaviour-guiding 'mentality' that neuroscientists attribute to awake, aware, healthy humans.

With these preliminaries and caveats, we can approach the linguistic minefield. In what follows, 'awareness' and 'consciousness' will be treated as synonyms; the word 'awareness' will be used preferentially as it has somewhat less connotative baggage. We cannot claim that the definitions given below would be accepted universally, but at least they are *reasonable* definitions, and they will be employed in what follows.

Phenomenal awareness: If I am experiencing X, I am phenomenally aware of X. If I am capable of experiencing X, then I am capable of being phenomenally aware of X: there is something that it is like, for me, to be aware of X. This is Block's phenomenal consciousness, the 'what it's like' or 'qualia' of experience. See Hacker (2002), however, for a warning that such phenomenal experiences may be effectively unique, may not usefully divide into 'kinds' and hence may not usefully be assigned ontological status or treated as 'entities' in a theory. The 'illusionism' of Dennett (1991) or Frankish (2016) is based on this view that qualia are ill-defined and hence inappropriate ontological bases for theory-building. Dennett (2017), echoing Lucas, rejects the claim that it is 'like' anything to be phenomenally aware of X without some other, meta-level awareness of being aware of X. This addition of a post facto criterion for awareness – reportability, for example – illustrates how the terminology in this area becomes a minefield. Note, moreover, that this is a *definitional* dispute, not one where evidence of any kind plays a role.

Non-phenomenal (or 'subliminal') awareness: Access consciousness of X without phenomenal awareness of X, as postulated for Damasio's protoself. The primary example of non-phenomenal consciousness in humans is the postulated role of the protoself: the constant monitoring of bodily functions that assures maintenance of homeostasis and hence organismal survival. Breathing, heartbeat, and the distribution of blood throughout the body, for example, are constantly monitored and adjusted as needs change, e.g. as one switches between difficult cognition, physical exercise and rest. All major organ systems, including the immune system, are similarly monitored and adjusted. The brainstem does all of this without conscious assistance, whether we are awake or in deep sleep. Most people are typically unaware, for example, of their blood pressure, liver function, or immune status, and are surprised when tests show either a dysfunctional or pathological state. 'Consciousness' is often contrasted with either anaesthesia or deep sleep. This contrast ignores non-phenomenal awareness. Would you agree to open-heart surgery while asleep? Or would you demand anaesthesia?

As requirements for autonomy, and particularly autonomous agency, for robots have come to the fore over the past two decades, the need to maintain homeostasis has increasingly become seen as the basis for making the robot's world and hence its tasks *matter* to it (Froese and Ziemke, 2009; cf. the discussion of mattering in The Third War). Homeostasis, in organisms, involves metabolism, the constant process of rebuilding and repairing the body using parts and energy obtained from the environment (technically, 'allostasis'). Such processes of self-maintenance correspond largely to the functions of the protoself; hence by analogy to humans, they may be expected not to require phenomenal consciousness. Where the maintenance of homeostasis *does* require phenomenal consciousness in humans is in the acquisition of materials and energy, e.g. food and water, from the environment. Types of phenomenal experience required for this task are discussed below.

A second well-studied example of non-phenomenal consciousness in humans is *priming*: the use of a briefly presented stimulus, of which a subject is not reportably phenomenally aware, to influence the processing of a later stimulus of which the subject is reportably phenomenally aware (see Kahneman, 2011 for extensive discussion and examples). Priming can be studied precisely because it is non-random; indeed it is predictable (e.g. by advertisers) and often rational.

With this distinction between phenomenal and non-phenomenal awareness, it is possible to formulate the question of whether self-driving cars can see as the question of whether they have phenomenal awareness or only non-phenomenal awareness. It is clear, after all, that they at least usually respond appropriately (i.e. rationally by our lights) to the presence of obstacles. Perhaps, for self-driving cars, everything is a prime (a case of priming, see above), but nothing has qualia? Does this mean a self-driving car has a protoself? Is this also a way to explain intelligent behaviour in micro-organisms, plants, and invertebrate animals?

Note, however, that this definition of non-phenomenal consciousness rests on a third-person criterion: reportability. Experimenters conclude that a subject in an experiment was not phenomenally aware of a priming stimulus when they insist that they did not see it (hear it, smell it, touch it, etc.). Is this conclusion justified? Might they have seen it, ever so briefly, but not remember seeing it? Primes are designed to be short – less than 100 ms (milliseconds: one thousandth of a second) – less time than is required for perceived scenes to 'make it to consciousness' (about 270 ms; Sergent, Baillet and Dehaene, 2005). This measured time is, however, also third-person. That non-phenomenal awareness even exists in humans is, therefore, only a plausible inference from third-person data. Maybe we are fully aware of the process of maintaining homeostasis, but unable to report on it.

Bodily awareness, somatic awareness, basic emotions: Damasio's typology is focused on bodily awareness and the role of emotions in decision making. Mammalian brains devote substantial cortical real estate to somatosensory perception (the famed 'somatosensory homunculus' with huge hands, face and genitalia) and proprioception. Much of the brainstem and midbrain is devoted to detecting and regulating bodily state and converting information about it into felt emotions (Craig, 2002) as well as bodily sensations like hunger, thirst, pleasure, and pain. These sensations are strong motivators of behaviour. The role of emotion in all aspects of cognition, including what used to be considered 'pure' thought, is now empirically well-supported (e.g. Kahneman, 2011; for specific roles of gut-brain interactions, see Mayer, 2011, for heart-brain interactions see Thayer and Lane, 2009).

As discussed in The Third War and Chapters 8 and 9 on ethics and ethical AI, emotion was mostly ignored during the first five decades of AI, i.e. during the AI Wars of Part I. With the rise of robotics, however, the kinds of somatosensory and proprioceptive feedback that underpin emotions in humans could no longer be ignored. Intrinsic motivation is now an essential research area within developmental robotics (Oudeyer and Kaplan, 2007; for a history of this transition in research priorities, see Picard, 2010).

As noted above, Damasio's protoself is defined as having non-phenomenal bodily awareness. Much of bodily awareness, including much of proprioceptive awareness, does indeed seem non-phenomenal (and is in fact non-reportable), at least much of the time. One is typically not aware of one's heartbeat or the state of one's stomach, intestines, or liver, though one can be. One typically does not know the positions of one's knees while walking, even though the proprioceptive and motor planning systems monitor and adjust them continuously. The notion of an unconscious pain seems like an oxymoron, but unconscious (or 'repressed') psychological pain is a mainstay of psychotherapy. Do unconscious pains exist? Again, they are an inference from third-person data.

Perceptual awareness: 'Percepts' are not just scenes (or their analogues for other senses), but rather interpreted layouts of categorized and sometimes individually identified objects, in an interpreted or at least interpretable event-like context, against a perceived 'background' that is taken to be irrelevant or minimally relevant to what is going on. Percepts, in other words, are heavily laden with semantics. How this works in humans is under intensive investigation and remains far from clear (Fields, 2016). Replicating this ability in a development-robotics context is an active area of research (e.g. Lyubova, Ivaldi and Filliat, 2016).

Perception, on this definition, relies heavily on conception: the ability to recognize and categorize objects (Fields, 2012). Hence the richness of perception depends on the richness of the categorization system. Categorization and hence semantic interpretation falls under Panksepp's secondary-process consciousness. It is worth noting that such categorization, including the 'perception' of causation, animacy, and agency, is involuntary for humans: one cannot 'choose not to see' causation, animacy, or agency just as one cannot choose not to see a dog as a dog or a Necker cube as three-dimensional (Gao, McCarthy and Scholl, 2010; Firestone and Scholl, 2016).

Perceptions may be accompanied, in humans, by a number of 'epistemic' feelings, including feelings of familiarity or unfamiliarity, puzzlement about whether a person or object is who or what you suspect it may be, the slight Aha! of successful but delayed identification (e.g. Eichenbaum, Yonelinas and Ranganath, 2007). Epistemic feelings may or may not be available to phenomenal awareness, but they can be, especially when expectations are violated. Such feelings are also typically available post facto: if asked, a person can typically report whether they have had a perceptual experience or an imaginative experience (see below). The kinds of experiences that count as 'perception' can differ widely between individuals; for example, synesthetes see sounds or feel tastes tactically in ways that non-synesthetes can scarcely imagine (Hubbard and Ramachandran, 2005; Ward, 2013).

Feelings of familiarity can specifically fail, e.g. in Capgras syndrome where the feeling of familiarity associated with close relatives fails and such people appear, to the patient, to be imposters (Hirstein and Ramachandran, 1997). A crucial such feeling is the 'existential' feeling of reality, that what is perceived is really there. This feeling is generated by a 'reality-monitoring network' that can fail dramatically, e.g. in psychosis (Simons et al., 2008; Griffin and Fletcher, 2017). The feeling of reality is often below the level of phenomenal awareness unless expectations are violated.

Imaginative awareness: Imaginations are modal, i.e. visual, aural, tactile, olfactory or gustatory sensory experiences, alone or in any combination, that are not perceptions. Imaginative awareness may be accompanied by a reality-monitoring feeling indicating that it is imaginative awareness, not perceptual awareness; however, an experience may also be identified as an imagination only post facto.

Episodic memory and imaginative planning are canonical examples of imaginative awareness, as are silent use of language, songs running through one's head, visualizations, dreams, hallucinations, and any other non-perceptual sensory experiences. Human imagination involves top-down activation of many of the same neurocognitive networks involved in perception (Kosslyn, Thompson, and Ganis, 2006). The initiation and flow of imagination is not well understood. While the mechanisms of episodic memory retrieval, for example, have been intensively investigated (e.g. Rugg and Viberg, 2013), how a particular memory is selected for retrieval remains unknown. While many mammals appear to dream, whether non-human animals experience episodic memories or other kinds of imaginative awareness remains largely uninvestigated.

Episodic memories typically 'feel like' memories; this allows one to 'know that one is remembering'. Retrieving a fact, e.g. the date of a friend's birthday, from 'semantic' memory similarly feels like retrieving something remembered, not perceiving it. Imaginative awareness is also often accompanied by bodily sensations and emotions, which may be as strong as those accompanying perceptions; a memory, for example, may evoke fear or disgust in the present.

What is commonly called 'thinking' is a form of imaginative awareness, typically a flow of words, images, musical sounds, or some combination of these. Memory of some kind for past events is clearly required for inferences from past experiences and for planning future responses to similar situations. What, however, is the advantage, assuming there is one, of *experiencing* such thoughts or memories? It is known, for example, that re-remembering or 'reconsolidating' episodic memories changes their content (Schwabe, Nader and Pruessner, 2014) and that the contents of episodic memories are, in general, easily manipulable by various means (Henkel and Carbuto, 2008). Would this still be the case if the memories were not experienced? Are there episodic memories that influence behaviour (and therefore have access consciousness) but are never experienced? Are such memories, if they exist, 'repressed' or merely non-phenomenal? Similar questions can be asked about experiences of thinking, planning and other uses of the imagination.

Autobiographical consciousness: An important special case of imaginative awareness is imaginative awareness of the 'self' as an actor and/or observer in both episodic memories and plans. Episodic memory in particular, is typically defined as self-referential, i.e. as involving 'autonoetic' awareness (Klein, 2014), although this felt sense of 'ownership' of episodic memories can be lost (Klein and Nichols, 2012). Humans display a high degree of variation in the ability to access autobiographical information via episodic memories, with some being able to imaginatively re-experience a wide range of verifiable events from both their recent and distant past and others almost completely lacking this ability (Palombo, Sheldon and Levine, 2018). Individuals who do not imaginatively re-experience their pasts via episodic memories may nonetheless have robust factual (i.e. 'semantic') memories of their pasts; however, these lack both the vividness and the emotional overtones of episodic memories. Both kinds of autobiographical memory can be lost, e.g. in dementia or Korsakoff syndrome (Fama, Pitel and Sullivan, 2012). Again, similar remarks can be made about planning.

Self consciousness and introspective awareness: The self appears as an actor and/or observer not just in the experienced past of autobiographical memories and the experienced future of personal plans, but in the experienced present. Crucially, however, this 'psychological' as opposed to bodily self does not appear in all of the experienced present. When attention is highly focused on perception of and action in the external world, even on *imagined* action in the external world, or on imagined action in an external but *virtual* world (e.g. when proving a mathematical theorem) the self tends to 'disappear' (e.g. Csikszentmihályi, 1990). Phenomenal awareness of the self correlates with activity in the 'default-mode network' (DMN), so called because it is active typically when humans are not engaged in an externally oriented, attention-demanding task (Buckner et al., 2008; Qin and Northoff, 2011). DMN activity correlates with self-relevant social emotions, self-evaluation in relation to others, and explicit metacognitive reflection on desires, goals and beliefs (Northoff et al., 2006; Schilbach et al., 2008). High levels of DMN activity correlate with both obsessive self-relevant ruminations in major depression (Sheline et al., 2009) and self-relevant delusions in schizophrenia (Kim et al., 2009).

The social and psychological self implemented by the DMN is functionally linked, in humans, with the bodily self, the sensory self and the reward system by a deep-brain association centre: the insular cortex (Craig, 2009). Severe disruption of these connections can result in specific loss of the 'existential' feeling that one exists as a living being, e.g. in Cotard's syndrome (Debruyne et al., 2009).

Introspective awareness is typically accompanied by epistemic feelings, including feelings of knowing or believing, certainty, doubt, or subjective probability, or of understanding or confusion. It can also involve autonoetic feelings, e.g. feelings of ownership of knowledge, beliefs, or desires, feelings of agency associated with actions or plans, and reality-monitoring feelings that distinguish these 'private' internal experiences from perceptions of 'public' events in the external world. Epistemic and autonoetic feelings of this kind do not reach awareness when attention is highly focused on externally oriented activity, i.e. when the reality-monitoring system is indicating perceptual interaction with the external world. One does not, for example, reflect on one's beliefs about how cars work or traffic laws while weaving through traffic at high speed.

As noted earlier, 'consciousness' is often used, particularly by philosophers, to mean the entire package of bodily, emotional, perceptual, and imaginative consciousness accompanied by introspective self-consciousness. One of the most important contributions of recent neuroscience to debates about consciousness has been to tease these various aspects of consciousness apart and to show, both experimentally and by analysis of various pathological states, that they can occur in different combinations. A person may be conscious, medically and legally, without being conscious in the 'robust' sense assumed by many discussions of consciousness!

Spiritual consciousness: A variety of strong emotions, including awe, wonder, ecstatic joy, existential and moral anxiety, and deep contentment may occur spontaneously or in association with a wide range of behavioural and/or neurocognitive correlates including intense physical activity, meditation, religious practices, ingestion of hallucinogens/entheogens (Masters and Houston, 1966; Griffiths et al., 2006), and insular-cortex seizures (Gschwind and Picard, 2016). Such experiences are often accompanied by epistemic feelings, particularly of 'knowing' or certainty, and autonoetic feelings, such as feelings of 'oneness' or of the absence of a self. It is now known that the canonical hallucinogens psilocybin and LSD enhance sensation while strongly decreasing DMN activity (Carhart-Harris et al., 2012; 2016). Common meditation practices also strongly decrease DMN activity (Brewer et al., 2011; Tang, Hölzel and Posner, 2015), suggesting that decreased DMN activity may be a common element of all such experiences (but see Josipovic, 2014 for evidence that some forms of meditation decrease the usual anticorrelation in activity between DMN and the external-task network). The role of insular cortex as a global integrator of bodily, emotional, epistemic and autonoetic awareness and its apparently specific role in representing the self as located in the present (Craig, 2010; Picard and Kurth, 2014) also suggest a common role in such experiences.

4. Who's Afraid of Automaticity?

With this background in hand, we can ask: what are we conscious of? The answer, clearly, is bodily sensations, emotions, percepts, imaginations of various kinds, the self with its beliefs, emotions and feelings, sometimes ecstasy or deep contentment. Which combination of these we typically experience is highly variable between individuals; some people, for example, almost never experience inner speech, while others almost never experience visual imagination and still others rarely experience perceiving, with attention, the external world (Heavey and Hurlburt, 2008). Evidence from pathology, including various syndromes mentioned above, indicates that many common forms of experience can be specifically eliminated without apparent effects on other kinds of experience. For example, memory can be eliminated without affecting perception, ownership of memories can be eliminated without affecting memory content, and even the feeling that one exists can be eliminated without eliminating other forms of consciousness.

A more interesting question for the present purposes, however, is the opposite of this one: what are we *not* conscious of? Everyone reading this chapter is very good at a number of things that they are seldom conscious of doing. Each of us, for example, is accomplished at the task of navigating a cluttered environment without running into things, not just on foot, but while riding a bicycle or a motorcycle or driving a car. How often do we pause and notice: 'I've just made it across this room without running into the furniture?' How often do we think: 'Wow! I can drive through traffic!' Not often one hopes, as doing so risks a fatal diversion of attention. This is, of course, not to say that we are not conscious – not aware – of the furniture or the other cars. It is the activity of navigating around these things that we are not aware of.

Each of us is very good at reading English sentences. Do we pause and reflect: 'I'm reading this sentence!' How often do we stop a conversation to think: 'I'm using English grammar to construct these utterances?'

The ability to engage in complex, attention-demanding cognitive processes 'without having to think about it' is typically termed 'automaticity' (Bargh and Ferguson, 2000; Bargh et al., 2012). Human infants and children rapidly achieve automaticity in a huge range of skills, from locomotion to language use, causal reasoning and 'theory of mind' (ToM) judgments about what other people believe or intend. Adolescents achieve automaticity in technical, athletic and social skills, from javascript programming to flirting. Adults achieve automaticity in professional skills, and often in difficult tasks such as second-language learning. Automaticity is a mark of *expertise*: if doing something just comes naturally, it is something at which you are an expert, even if you make characteristic and possibly stereotypical errors.

Four features of automaticity are of particular interest here. First, walking across a room, driving a car, speaking a natural language, dancing, painting, playing tennis, or delivering a lecture require being *conscious*. In particular, these activities require, with odd exceptions like sleepwalking, *phenomenal awareness* and specifically, bodily awareness and perceptual awareness. To test this, try walking across the room with your eyes closed. Even if you are performing some action on 'full automatic' you are *not a zombie*. If you were, you would be incapable of your performance. 'Automatic' performances cleanly dissociate consciousness from meta- or self-consciousness. You can be fully conscious even in the complete absence of meta- or self-consciousness, as robustly documented for the kinds of externally (i.e. non-self) focused experiences, often called 'flow' experiences, described earlier (Csikszentmihályi, 1990).

Second, virtually everyone achieves automaticity in some number of non-overt, purely imaginative skills. You may, for example, talk yourself through the solution of a problem or formulation of a plan fully automatically, without stopping to think: 'I'm thinking!' You may visually imagine an artwork, compose music, work out the steps of an algorithm, or plan a route up a difficult rock face with the automaticity that comes with expertise. In such cases, you are engaging imaginative awareness without meta- or self-consciousness. Automaticity does not require external perception and is not just a matter of well-rehearsed behavioural skills. The great variety of abilities that humans can and regularly do 'automate' is the best evidence to date that Fodor's distinction between 'modules' for specific tasks and a 'central processor' for everything else is an enormous oversimplification (see Supplement 6.1). Humans are not vast collections of monointelligences for driving, rock climbing, composing symphonies, performing ballet, and hundreds of other activities. But these activities are, at least in experts, fully 'automated' and performed without self-consciousness.

Third, *noticing* that you are performing an action 'on automatic' requires a pause during which your attention shifts from *doing* whatever you are doing to *realizing* that you are doing it. *Introspective* awareness interrupts automatic performance. This is reflected in idiom: if a performance is described as 'self-conscious' we can infer that it was not expert. It takes an expert actor, for example, to *feign* self-consciousness convincingly.

Abilities learned later than the preschool years typically require a period of self-conscious practice before automaticity is achieved. Driving a car, for example, does not 'come naturally' in the way walking or speaking a language does; it must be learned through instruction and careful practice in relative isolation from the demands of fluid, expert performance. Introspective awareness in the form of self-conscious rehearsal is required during this learning phase but drops away as expertise develops. It *has to* drop away; pauses for introspection at high speed cause car wrecks.

Fourth, it may be difficult or impossible to say *how* we perform activities that we perform with automaticity. Explain, for example, how you walk or drive a car. The failure of protocol-driven knowledge engineering is commonly acknowledged as one of the principle failures of GOFAI as discussed in the Second War (in Chapter 4). Decades of research have failed, thus far, to explain how we formulate grammatical sentences or understand the meanings of sentences that we hear. It is not just difficult for *us* to explain our expert performances; it is difficult for science to explain them. Our expert performances are hard to reverse engineer, and introspective awareness provides, in case after case, no help whatsoever.

These features of automaticity naturally raise a question: what would it be like to be expert at everything? What would it be like, for example, to learn all abilities the way we learn to walk and talk, with exposure to expert performances by others, practice, and corrective feedback from the world, but without self-conscious practice requiring introspective awareness? What would it be like to display expert performance without *noticing* that you were performing or *noticing* that your performance was expert? What would it be like to correct errors without *noticing* the error or *noticing* the correction, the way you learned as an infant to balance on two legs or learned in early childhood to form English sentences with subject-verb-object construction? What would it be like to engage in inner speech or visual, aural or tactile imagery without *noticing* that you were talking to yourself or visually planning a route to a friend's house? What would it be like, in short, to have robust bodily, perceptual, and imaginative awareness but no introspective awareness? What would it be like to engage in robust cognition but no *conscious* metacognition?

We are in fact surrounded by systems that nearly all evidence suggests live in precisely this way: other animals. Animals hunt, navigate, mate, defend themselves, and engage in various social behaviours with considerable expertise. They display every possible sign of bodily and perceptual awareness and in many cases appear capable of planning, though whether they have robust imaginative awareness remains controversial (Suddendorf and Corballis, 2007; Penn, Holyoak, and Povinelli, 2008; Gentner, 2010).

They do not, however, appear to have introspective awareness. They do not, at least to all appearances, reflect on, mull over, criticize, or even *notice* their own expertise.

Dennett (2017) has described cognition without introspective awareness as 'competence without comprehension'. It is competence in the eyes of the world: successful performance, in other words. It is clearly lacking in comprehension of *how* the competence is produced, as discussed above. Does it, however, lack comprehension of the domain in which expertise is displayed? Dennett follows Searle's description of the Chinese Room in adopting a post facto criterion of comprehension: the ability to provide *reasons*, in the sense of justifications, for an action. A student, for example, is regarded as comprehending a mathematical proof if she can supply a reason for each step in the proof. Indeed, Dennett argues that the 'game' of demanding and supplying post hoc justifications was one of the primary drivers of the evolution of both language and comprehension (see particularly Dennett, 2017, Ch. 13).

Providing reasons for actions appears to be a canonical metacognitive activity: I can see (now) that I did X (in the past) because of Y. In such cases Y can be a fact or a belief, but also can be an intention or desire. As Dennett argues, providing reasons involves reverse-engineering one's own actions using introspectively available evidence. An offered reason only counts as a *good* reason, and hence only counts as evidence of comprehension, if those hearing it agree that it justifies the action, i.e. if they can imagine it justifying the action for them. This suggests, however, that comprehension is not a fact, but rather a third-party judgment. I may *think,* introspectively, that I comprehend, but be wrong in the eyes of others. At exam time, it is the eyes of others that count.

With these thoughts about comprehension, let us revisit the Chinese Room. Suppose the room has monitors that allow the diligent symbol-manipulator inside to see and hear the interlocutors outside. Suppose these interlocutors give every indication, in expression, gesture, tone of voice, etc. that the ongoing conversation makes perfect sense. Seeing and hearing this provides evidence of competence to the person in the room. Might this evidence not also be evidence of comprehension, even though it conflicts with the symbol-manipulator's introspection-based feeling of incomprehension? Consider a less far-fetched example: engaging in conversation in a second language in which one is far from fluent. One may experience oneself as just guessing at meanings, and just guessing at responses. How one's interlocutors react, in this case, is the best evidence one has as to both competence and comprehension.

It is possible to think, in such a case, 'I clearly don't understand'. But it is also possible to realize, 'I do understand this!' when the responses of one's conversation partners indicate success. Either way, the telling evidence is all third-person: introspection has nothing to do with it.

We can now ask: what makes introspection special? What is the difference between obtaining a reason for an action by conscious introspection and asking (silently or aloud) 'Why did I do X?' and waiting to hear an answer? What would it be like if *all* of our metacognition was unconscious, merely automated? What would it be like if we were metacognitive experts, knowing why we did things 'without having to think about it'? What would it be like to give reasons for actions without noticing any difference – indeed, without there *being* any difference, for you – between giving a reason and reporting any other kind of fact?

5. One Kind of Thinking or Two?

It is a commonplace of cognitive psychology that there are two kinds of thinking, often called 'process (or type or system) 1' thinking and 'process (or type or system) 2' thinking (see Evans, 2003; Kahneman, 2011 for reviews; Evans and Stanovich, 2013 for responses to criticisms; see Melnikoff and Bargh, 2018, and Mercier and Sperber, 2017 for further criticism). Type 1 thinking is fast, intuitive, heuristic, biased and automated. Type 2 thinking is slow, deliberate, rational, rule-based and conscious. Type 2 thinking is, in particular, introspective and involves giving reasons for each step in the thought process to oneself. It is what we commonly call 'thinking' and what we do when someone tells us to 'think about it' before answering or acting.

Using this language, we can rephrase the question about automaticity considered above as 'what would it be like if we only did Type 1 thinking?' We can also ask the opposite: 'what would it be like if we only did Type 2 thinking?' This latter question has a ready answer: it would be like careful, logical, well-justified inference using well-defined symbols and drawing on a self-consistent network of facts, each labelled by its relevance to various possible situations and probability of being true. If we only did Type 2 thinking, we would be physical symbol systems, as Newell (1980) defined them. Our mental life would be just like executing a GOFAI program. Indeed GOFAI can be defined as AI that assumes that what we experience as Type 2 thinking, stripped of its errors, oversights, and inconsistencies, is what thinking *is*. Descartes might well have agreed.

Implementing a GOFAI program on a conventional, serial von Neumann machine (a standard computer) is challenging but conceptually straightforward: von Neumann machines are designed to perform sequences of well-defined computational steps on precisely specified symbolic representations. How would we implement such a program on something more like the brain, with a Bayesian-network (Friston, 2010; Clark, 2013) or even a standard deep-learning (LeCun, Bengio, and Hinton, 2015) architecture? Fortunately we have an example answer to this question: AlphaGo (Silver et al., 2016). The 'symbols' for AlphaGo are board positions; the 'steps' are predicted next+1 positions following candidate moves computed by a neural network trained by deep learning from tens of millions of Go moves. We can expect this answer to generalize: the way to get a brain to execute long sequences of 'rational' inferences is to give it a massive experience-base of successful and unsuccessful inferences from which to learn. The criterion of 'rationality' in this case is not theoretical conformation to some idealized model, but rather real-world success. The multi-step inferential process is rational, post facto, if it *works*.

This generalization from one example is surprisingly close to a well-supported, high-level model of human neurocognitive architecture: the 'global neuronal workspace' (GNW) model (Baars, 2005; Baars, Franklin, and Ramsoy, 2013; Dehaene and Naccache, 2001; Dehaene, Charles, King, and Marti, 2014). According to the GNW model, the brain is a relatively sparsely connected network (i.e. a small-world or 'rich club' network; see Sporns and Honey, 2006; Shanahan, 2012) of relatively richly connected subnetworks. The high-level, sparsely connected network is the GNW; the lower-level networks it connects perform relatively specialized functions such as visual or aural scene analysis, emotional response, planning, or motor control. High-amplitude, persistent, oscillatory activity in the GNW is experienced as 'events' or 'states' of both the body and the (external or imagined) world. Note that while the GNW model is anatomically, functionally, and correlationally well-supported, its identification of GNW states with experience remains an arguably nondemonstrable hypothesis (see The Third War, Supplement 5.5).

The GNW model makes two significant additions to the generalization from AlphaGo outlined above. First, the GNW is a network of specialized subnetworks, the distinct architectures of which are specified genetically and largely developed prenatally (Ball et al., 2014; Gao et al., 2015; Huang et al., 2015). Experience-based learning does not, therefore, have to configure these networks from scratch: it has a substantial evolutionary head start. Second, the GNW model includes the networks that generate emotions and epistemic, agential, and existential feelings. Thus GNW 'states' include the 'markers' that indicate whether experiences are perceptions or imaginations, memories or goals, pleasant or noxious. While AI tended to ignore such things in the pre-robotics era as noted above, neuroscience has not; indeed understanding how these feelings are disrupted or altered in pathological or altered states of consciousness has been a significant goal of research.

The AlphaGo, GNW 'picture' of cognition sketched here suggests a fairly radical hypothesis about experience: that cognitive/emotional/feeling *states* are experienced, but the *processes* generating these states are not. If this is correct, Type 2 thinking is just Type 1 thinking with lots of intermediate steps. Each 'step' between experienced states, in particular, is an instance of Type 1 cognition. Deliberation, in other words, is just a sequence of small intuitions. Type 2 cognition is more rational than Type 1 cognition because the steps are small and the intermediate states are represented explicitly: solving a problem using Type 2 cognition is, as typically the case in logic or mathematics, a matter of reformulating it over and over until a formulation is reached for which the solution is obvious and immediate.

A particularly clear formulation of this idea has recently been provided by Chater (2018, p. 180, emphasis in original): 'mental processes are *always* unconscious – consciousness reports answers, but not their origins . . . we are only ever conscious of the results of the brain's interpretations – not the "raw" information it makes sense of, or the intervening inferences'. Chater's 'flat' conception of the mind comports well with a formal model of a 'conscious agent' (CA) as a system capable of perception, decision making and action (Hoffman and Prakash, 2014; see Fields, Hoffman, Prakash and Singh, 2018 for an extension of this model to agents with memory, emotion and planning ability). A CA is characterized by a set of possible experiences. This set may be very large, but is finite provided 1) the total resolution of the agent's experiences, across all modalities, external and internal, is finite, and 2) the agent survives for finite time. Other than the experiences in this set, a CA by definition experiences nothing. It does not, in particular, experience its own decision processes or actions, nor does it have direct experiential access to the 'world' it inhabits.

Perception and its synthetic analogue, imagination, provide a model of the structure of the 'answers' that we as conscious agents experience. They are, as noted earlier, not merely scenes in some modality or combination of modalities. They instead comprise entities – distinct visual objects, melodies or phonemes, textures, tastes, odours – embedded in a background that is featured but not further divided into recognized entities and structured into a spatiotemporal 'container' in which the distinguished entities are localized. Other components of experience – pain and pleasure, attraction and revulsion, feelings of knowing, acting, and ownership – are overlaid on these highly structured layouts of entities and background, with our own bodies typically being prominent among the entities involved. These components of experience, taken together, constitute the *umwelt* (von Uexküll, 1957) or 'manifest image' (Sellars, 1962) in which we live (what we called a cognitive world in Chapter 5). We reify them and consider them to provide an ontology, a catalogue of what exists (Dennett, 2017). This experience-derived ontology is specific to us humans, and indeed specific to each of us individually. We can communicate effectively to the extent that we inhabit and largely share each other's ontologies; as our languages and cultures become more distant, communication breaks down. We can scarcely imagine, in Nagel's (1974) famous phrase, what it is like to be a bat, and what it is like to be a tree or a bacterium escapes us altogether.

Where did this ontology of objects, events, words, emotions, and feelings come from? Within the current scientific worldview, the only possible answer is evolution. Hoffman and colleagues have shown, by a combination of evolutionary game-theory experiments and formal analysis, that if our ontologies are the result of a natural selection process, then what they encode are fitness opportunities and costs (Hoffman, Singh and Prakash, 2015; Hoffman, 2016). The space-time 'container' in which the elements of our ontologies are laid out is effectively an interface; the objects, events, words, emotions, and feelings displayed on that interface are icons summarizing fitness opportunities and costs of possible decisions and actions. This 'interface theory of perception' (ITP) comports well with over a half-century of cybernetics and information theory, and with the emerging consensus among physicists that space-time is not fundamental, but rather emergent from lower-level, probably informational processes (Fields, Hoffman, Prakash and Prentner, 2017). Whatever the world is 'really' like bears, in all likelihood, no relationship at all to the practical ontology of our more-or-less shared *umwelt* other than one: it offers the opportunities and enforces the consequences that have made us, and our *umwelt*, what they are.

6. Flat Phenomenology

Understanding experience in terms of ITP, or in terms of Chater's 'flat' conception of the mind, destroys naive realism about both 'external' and 'internal' worlds. What we iconify as tables and chairs, other people, and even our own bodies are aspects of a complex fitness landscape in which we, as experiencers, exist. What we iconify as beliefs, desires, intentions, and emotions are aspects of the same fitness landscape. As Chater puts it, we have no unconscious minds, no store of propositions of which we are not currently aware, but to which we have (unconscious) attitudes such as knowledge, belief, subjective probability, desire, or moral feelings. We have no stores of recorded experiences or written-out facts in memory. There is no 'language of thought' (Fodor, 1975). What we have instead is a question-answering process that *makes up* answers to questions about what we know, believe, or feel, moment to moment in response to whatever situation we are in. The primary criterion on these answers is coherence with other answers delivered recently, including answers about what we are currently seeing or hearing, who we are with, etc. Our question answerers, in other words, work like neural networks, constantly adjusting their connection weights in response to the totality of their ongoing experience. There are no 'hard constraints' that force some connections to remain fixed. There is no standard of consistency that we must or even can meet.

Like naive objectivity, ITP and the 'flat' conception destroy naive subjectivity. There is no longer a 'subject' of experience; there is just experience. The experienced 'self' is an icon; it participates in some experiences (the introspective ones) but not others. 'Losing one's self' in a 'flow' experience (Csikszentmihályi, 1990) or in meditation is not at all surprising on this view; these are simply experiences in which the 'self' icon is not activated. The only reason the 'self' icon would be activated is if fitness opportunities or consequences that it encodes are relevant to a particular action.

The iconified 'self' does not experience anything, does not have any mental states, and does not control what is going on (cf. the 'illusory' or 'constructed' self of Metzinger, 2003). Neither, of course, does an iconified 'body' or an iconified 'brain'. These icons are associated, on the interface, with other icons representing other kinds of opportunities or consequences. They are components of experience, not objects or subjects of experience.

It is interesting in this regard that ruminative focus on the iconic self, its history, and its social relationships broadly correlates with low self-reported well-being and/or clinical depression (e.g. Nolen-Hoeksema, 2000; Harrington and Loffredo, 2011). Consistent with its major role in introspective experience, hyperactivation of and reduced ability to deactivate the DMN correlates with ruminative thinking and depression (Sheline et al., 2009; Hamilton et al., 2011). Hence, far from an unalloyed gift from nature, the human iconic self and the introspective awareness that examines it can be a major source of suffering.

7. What is the 'Consciousness War' About?

The question of consciousness lurked just below the surface throughout the AI wars of the twentieth century's last five decades. Turing, after all, brought up and deferred the issue of consciousness in his seminal paper on AI: 'I do not wish to give the impression that I think there is no mystery about consciousness ... But I do not think these mysteries necessarily need to be solved before we can answer the question with which we are concerned in this paper' (Turing, 1950, p. 447; see Chapter 2). Lucas (1961) was far less sanguine, proposing that an effectively infinitely capable, unified consciousness set humans apart from any possible machine. With the development of neuroimaging techniques and the consequent rise of cognitive neuroscience, the question of consciousness has come to the fore. The AI Wars have, in many respects, morphed into the Consciousness War. At stake is not just whether *machines* can be conscious, but whether *brains* or even *organisms* can be conscious.

The starting point for all sides in the Consciousness War is that *we* are conscious. We are, each of us, conscious by ostensive definition: this having-of-experiences that we escape only in deep sleep or anaesthesia is what we mean by consciousness. We assume that each other are conscious because in point of fact we must; from a more philosophical standpoint, we do so because not doing so would be uncharitable. The questions, then, become 1) what kinds of entities *we* are, and 2) what *other* entities are sufficiently similar to us to count as conscious? As we have seen, prevailing academic opinion has admitted animals and even human infants to the 'consciousness club' only recently. If we are reluctant to admit that even our own offspring are conscious, what chance is there for machines?

Do we really care whether self-driving cars can see? Many claim to, but we suspect a deeper question is what drives their concern. Whether it counts as 'seeing' or not, we know how self-driving cars avoid obstacles. We do not, perhaps, *want to know* how we do it. The idea that systems *we built* might be conscious is a particularly brutal and uncompromising version of the idea that systems *we understand* might be conscious. We do not, one suspects, want our consciousness to be understood. We do not want a 'theory of consciousness' that explains how we work.

While some theorists, notably Tononi and colleagues, have sought to *define* consciousness (Tononi, 2008; Tononi and Koch, 2015), the science that has emerged over the past two decades has largely sought merely to characterize the contents of consciousness in particular instances, to search for 'neural correlates' of this or that experience. This search has been devastatingly effective in showing that particular kinds of experience can be dissociated from other kinds. Human beings can perceive without remembering, act without knowing they are acting, think without recognizing their thoughts as their own, and go through life claiming that they do not exist. Cognitive neuroscience has also shown, with alarming clarity, that we cannot understand how our minds work by introspection. We cannot even trust our memories to tell us what happened yesterday.

Science has not, however, told us what consciousness is. It has not given us generally acknowledged, public, observational methods for determining whether an entity is conscious. It has not told us where to draw the line between conscious and unconscious entities, though such lines are constantly being proposed, debated, and rejected. What science has steadily done is, instead, to tell us what consciousness is *not*. It is not a uniform 'package deal' of capabilities. It is not a window onto the true structure of reality. It is not even a window onto the true structure of our own minds. It does not give us control over ourselves or the world; indeed it does not allow us to predict our own actions from one moment to the next. Science has told us, in no uncertain terms, that having 'clear and distinct ideas' is no guide to the truth.

Science is telling us, in short, that consciousness is not what we thought it was. Hence it is telling us, as Copernicus and Darwin told us, that we are not who or what we thought we were. Like the AI War that preceded it, the Consciousness War is not about self-driving cars. It is about us, and whether we are special.

8

Ethical Issues Surrounding AI Applications

1. Introduction

The great debates and discoveries of AI proceeded for nearly fifty years before anyone, outside of science fiction, spent much time thinking about the ethical impacts that these technologies might have on individuals and societies. Of course, there were a few notable exceptions. Alan Turing (1950; see Chapter 2) lists the 'heads in the sand' objection as the second of the nine objections to machines having 'intelligence' that he addresses. This objection to machine intelligence rests on the idea that the consequences of machines' thinking are just too terrible to conceive of, and so: 'Let us hope and believe that they [the machines] cannot do so [i.e. think]' (ibid.). Perhaps it would not be too much of a stretch to assume that this objection might also be extended to include the proposition that we should not build thinking machines even if we could technologically do so. Turing points out the obvious counterpoint – just because something seems unthinkable or undesirable does not mean that it will not come to pass. This is true, but it raises the question of why these machines strike some people as being a terrible idea. The fear comes with the intuition that AI machines will indelibly change the way of life humans have enjoyed for some time. Stephan Hawking, arguably one of the most intelligent humans to have lived, publicly announced on 20 October 2016, that he foresaw a day in which the study of AI would not only succeed in building intelligent machines but that it would also go on to create machines with intelligence that could vastly surpass that of any human, or even all humans put together. At that point:

> Success in creating effective AI, could be the biggest event in the history of our civilization. Or the worst. We just don't know. So we cannot know if we will be infinitely helped by AI, or ignored by it and side-lined, or conceivably destroyed by it.[1]

This is a frightening prognostication. At this same event Hawking was optimistic that if we 'employ best practices and effective management' then we would be able to mitigate the worst outcomes, but he gave no argument that this would be the likely outcome. Turing's only answer to those who fear the world with uncontrollable AI that Hawking predicted, but do not share Hawking's optimism that we will find a way to manage advanced AI technologies, is a dismissive one. He suggests that these fears are '... likely to be quite strong in intellectual people, since they value the power of thinking more highly than others and are more inclined to base their belief in the superiority of Man on this power' (ibid.). Turing's next comments are particularly dismissive:

> I do not think that this argument is sufficiently substantial to require refutation. Consolation would be more appropriate: perhaps this should be sought in the transmigration of souls.
>
> Ibid

Obliviously, Turing is not one who takes the fear of AI replacing humans very seriously; perhaps he might even welcome it in some way.

Turing also suggested in his famous paper that AI would succeed best if we developed machines that could learn and grow on their own in ways that are similar to how human children learn (see section 7 of Turing, 1950). We can see that these machines would not have biological limits and any limits they might have would be technological ones. Given that technological limits have historically been only temporary setbacks, it is conceivable that such a machine could grow almost indefinitely as each successive technological limit is overcome. It follows that Hawking's worry that we might create machines that grow beyond our control, which might then pursue their own alien goals, might arise to the detriment of humanity. The act of building technologies with the capacity to grow beyond control would hardly seem like an ethical choice. That is, unless the destruction of humanity itself was ethically justified. Strange as that conclusion may sound, one of the present authors (Dietrich, 2011) has argued that we are ourselves examples of intelligence run amok, leading to the unjustified extinction of thousands of species, many of them sentient. Therefore, the potential good that other species would receive by our absence from the planet would be a kind of ethical good, especially if the AI that replaces us is more capable of reasoning in an ethical manner than we humans are. For instance, vast AI intelligences could, and perhaps would likely, be much better at managing the complexities of environmental conservation and science and thus could help a maximum set of biological species to survive on this planet for much longer than they would naturally. This is just one of many outcomes that we have trouble assuring, but which our AI superiors might be able to accomplish. As Dietrich argues:

> Once we implement moral machines, it would be a triviality for them to generalize maximally. They would instantly complete the Copernican turn we have yet to complete over the last 200,000 years – the time *H. sapiens* has been on the planet. The moral machines would instantly be our moral superiors (it follows that they wouldn't kill us). Rather than taking thousands of years to figure out that slavery is wrong, they would know it is wrong the minute they considered the concept of slavery. And such machines would soon surpass us in intellectual power. They will do better mathematics, science, and art than we do (though the first two are universal, the third is not, so we may not care much for their art). On the available evidence, then, it should be possible to implement machines that are better than we. The moral component of being better than we saddles us with a moral duty to build such machines. And then, after building such a race of machines, perhaps we could exit with some dignity, and with the thought that we had finally done the best we could do.

Dietrich, 2011b

Yet it is hard to find evidence that the great minds that have debated the what and how of AI that we have pursued in great detail in our earlier chapters paid any formal attention to the social and ethical impacts that were sure to happen if they succeeded. When it comes to ethical thinking, most of the early AI researchers were either disinterested, incompetent, unimaginative, and/or possibly amoral.

2. AI Ethics and the New AI Wars

The examples above show that some thought was given to the ethical impacts of AI and robotics, but these explorations were not central to AI research. Given that AI research was slow to show significant results that could be easily used to build working products, it was easy for the early explorers in this domain to ignore the potential ethical impacts of their research. With the big technological applications of their work being still theoretical, thought about AI ethics was left to science fiction, since that genus of literature likes to extrapolate on the potential impacts of as-yet unrealized science and technology. AI researchers were in a position to be able to think only about their own narrow research interests, largely in isolation from outside influences, with only the other people in their lab having any input into the programs that they produced. These programs might leave the lab as demos or showpieces to give proof of concept, but mostly they would be shelved after a time and the researcher would move on to some other program. These intellectual explorations in computer code had little impact outside of the research lab that created them and certainly few social or ethical impacts. While an AI program that can always win or draw in a game like tic tac toe is a nice programing achievement that is often recreated by students learning to program AI, these early successes in game playing computers did not change the world beyond the humility they evoke when one has to concede that one is never going to beat the machine in this game. At the time, many felt that this augured the eventual dominance of AI in all manner of games, that it was just a matter of time.

Tic tac toe is a solved game where every move can feasibly be mapped out and a perfect strategy employed even by a human game player. These kinds of simple games were easy prey for early AI that could quickly search the tree of all possible moves that would result from the moves of one's opponent and then pick the move that would eventually result in a win or a draw. More complex games such as checkers, where the combination of all possible moves went well beyond the ability of this brute force game-tree search method to solve, forced programmers to come up with different programming strategies of playing the game with more algorithmic flexibility than a simple exhaustive search. Even so, this game was cracked by the 1960s (see Samuel

1963, 1967). But when it came to building machines that could play world class chess, many years and whole careers were spent discovering a way for AI algorithms to play chess so well that it could win against a chess master. Eventually, though, in 1997 the reigning chess master Gary Kasporav was beaten by an IBM-created AI program called Deep Blue. Other games have also fallen to the march of AI, including the game Go, which was once thought to have far too complex a game tree to ever be effectively climbed by a computer program. Yet this game too was won at the master level in 2016 by Alpha Go, a deep learning program developed by the company Deep Mind. (For more on what computers can win at, see Supplement 5.1).

Strategy game playing has been a major research wing of AI and has produced astonishing results, but at the end of the day it is about playing games, the winning and losing of which have few obvious ethical impacts. The human players that were defeated by these various machines have no doubt suffered some depression at their loss, and the news media has printed humbling headlines as the human dominance of human-created games has come to an end. But these events, even if they were a media splash, fade back into the background, and those few humans who make their livelihoods playing specific games like chess or Go continue. There are still people happily playing chess and Go even though they know that a machine could beat them. Since the AI that has been in the public eye was largely focused on games, it is natural that the ethics of AI was of little concern from the early days of AI until very recently. AI of that era focused on domains that had little social impact, even when AI achieved its most important goal of defeating master players of various intellectual strategy games.

But over time, these explorations of AI produced more and more applications that left the toy worlds of simulation and research and moved into the real world as serious applications that could have wider impact on the world we live in. Even this was a slow process. For instance, the three volume *Handbook of Artificial Intelligence* has large sections devoted to AI game playing, but it also lists language understanding, applications to science, medicine, education, machine learning, cognition, deduction, vision, and planning. All these areas have obvious ethical impacts but even so, there is no mention of AI ethics in any of the three volumes of this work. It would still be a decade or more before this aspect of AI began to receive the attention it deserved.

While AI ethics had become a regular topic in applied ethics by the 1990s, it is only the very recent successes of AI technologies (e.g. deep learning with sophisticated ANNs) that have gotten the attention of the rest of the world; now many see the necessity of addressing the ethical problems that AI technologies present. Today AI is an enabling technology at the centre of an industry engaged in collecting, analysing, and transmitting vast amounts of data. Data is king in advertising, and the ability to make accurate deductions, for instance about the next movie you might like to watch, requires sophisticated AI programming that accurately deduces a range of choices for you in your movie list. This ability AI exhibits, while useful at times, can be a bit concerning, especially when the same techniques are used to predict your shopping habits, your voting habits, and your potential to commit crimes or skip bail. AI powered consumer predictive technologies have swept the market, and this has now required researchers to become aware of the impact their work has on the world in ways that have political, legal, social, and ethical implications. This change has happened rapidly and continues to accelerate. As AI applications find their way into every aspect of our lives and society, the urgency of understanding the potential impacts of AI technologies becomes even more apparent. In the next section we will introduce some of the most important ethical impacts that have evolved as AI has left the ivory tower and swept across the world. Such ethical discussions have led to new AI wars – wars about the ethics of AI applications – and their results will determine the rights we will have (or not have) in the face of technologies that have all the data about us that exists and can use that data to predict the choices we are likely to make. It is one thing to have a computer accurately predict your next move in chess, it is quite another when it predicts how to best kill you on a battlefield, or the next candidate you are likely to vote for, and nudges you to make certain political decisions without your knowledge. The effects that AI algorithms can have on the lives of individuals can be quite damaging. It has been shown that AI algorithms used in predictive policing, image processing, mortgage calculations, searches, and the ways that AI systems categorize people along various dimensions has proven to be embarrassingly sexist, racist, and to regularly reinforce racial injustice (Benjamin, 2019; Noble, 2018). Black AI researcher Joy Buolamwini discovered a pernicious problem while trying to debug a facial tracking system she was working on, where the system was unable to track her face unless she covered it with an all-white mask. This inspired her to form the Algorithmic Justice League (https://www.ajl.org/) and Buolamwini and many other people of color have stepped forward to criticize the injustice that AI algorithms have been shown to propagate. This is a new and important development in the AI ethics wars that early researchers who built face tracking and predictive policing apps failed to consider until they were forced to by activists.

3. AI is Dead! Long Live the AI App

As we have seen in earlier chapters of this book, the initial motivation of AI to replicate and exceed human thinking through artificial general intelligence ran into serious conceptual difficulties. This seeming falter allowed a new generation of AI practitioners to arise who instead sought to apply specific AI tools and techniques, such as deep learning and data mining, to tackle real world problems and create useful and profitable applications.

Stanley Kubrick, the director, producer, and coauthor of the screenplay for the epic 1968 movie *2001: A Space Odyssey*, created a character in the form of an AI supercomputer named HAL 9000. On a trip from Earth to Jupiter, HAL runs the ship, plans the missions, and plays chess, all the while maintaining convivial conversations with the crew. Famously, Kubrick employed scientists and technologists to help him make as realistic a representation as he could of the year 2001. Included in that group were AI luminaries such as Marvin Minsky, who helped him imagine the coming of AI. Naturally, they imagined a computer that could beat humans at chess and run complex systems on the ship, and one that could pass the Turing Test and speak to the crew. Many of these prognostications were realized, but not the ability HAL has in applying intelligent thought across cognitive domains, communicating effectively with humans, and plotting against the crew to remove them from the ship.

In 1968, it seemed like a reasonable bet that AI would be able to do these kinds of things, but the reality has fallen short for the reasons outlined in the earlier chapters of this book. We can speak to our machines to trigger certain actions from them, but we don't have meaningful conversations and they certainly do not have any form of consciousness or personality that cares one way or another about whether they win a game of chess or complete a mission to Jupiter.

But even though human-like AI has not come to pass, nonetheless, two decades after the actual date 2001, what the philosophers called 'weak AI' is a large part of our everyday lives in ways that were not entirely predicted, and many of these have had tremendous impact on our ethical lives. Even though there are no conscious AI entities plotting the overthrow of the human race, AI applications have been deployed on the battlefield, in commerce, education, medicine, elder care, home automation, and entertainment. In the next sections, we will take a quick look at some of the major ways in which AI applications are already making us think about how they are changing the way we live our lives, sometimes for better, but sometimes for worse.

4. Actual AI wars: Semi-Autonomous and Autonomous Weapons[2]

AI is featured in more than just the figurative philosophical wars discussed throughout this volume, and its influence goes beyond the many science fiction cases where AI is depicted as some terminating force arrayed against the human heroes. AI has, in fact, a growing presence on real battlefields today, as first noted by Singer (2009). The use of robotic weapons systems is accelerating around the globe. While the date of the first introduction of telerobotic weapons to the battlefield is debatable, it is clear that they have grown from the use of guided missiles and radio-controlled bombs in the last century to the 'smart' missiles and unmanned weapons systems of today. These systems are constantly evolving and a vast array of telerobotic and semi-autonomous weapons systems are being deployed in all potential theatres of conflict, land, sea, air, space and cyberspace (Singer, 2009). The epochal change in conflict resolution represented by the rise of more capable and autonomous weapons systems has not come without criticism. (We note that the more autonomous a drone is, the more AI is has. Nowadays, almost all drones have some autonomy.)

If you are a politician in a liberal democracy, the technology of unmanned weapons is the answer to your dreams. While armed aggression between liberal democracies is rare, they are involved in many military conflicts driven by clashes with nondemocratic countries or interventions in unstable regions of the world. Given the norms and values espoused in liberal democracies, there is a political will to spread the virtues of democracy and check the aggressions of nondemocratic powers. But other values and norms, such as the distribution of political power to the voting population, severely hamper the governments of those countries that try to act on their military aspirations. There are massive political costs to be paid when it comes to deploying large numbers of troops in foreign lands. Sauer and Schörnig (2012) note that the governments of democracies want to engage in military activities, but the citizens of democracies demand that these adventures be low-cost with no casualties from their own military and low casualties inflicted on the enemy and local population. They write:

> [T]he need to reduce costs, the short-term satisfaction of particular 'risk-transfer rules' for avoiding casualties, and the upkeep of a specific set of

normative values – constitute the special appeal of unmanned systems to democracies.

Sauer and Schörnig, 2012, p. 365

Unmanned weapons systems would seem to allow all constituents within a liberal democracy to achieve their goals. The weapons are not expensive compared to the massive expense of building, deploying and maintaining manned systems. The missions are secret and so small they hardly warrant mention in the daily news back home. There is almost no risk to military personnel in their use and the number of enemy casualties per strike is relatively small. Also, politicians such as President Barack Obama have made claims that these weapons are far less indiscriminate and more tightly controlled in their killing than alternative modes of attack (Landler, 2012). We should note that this claim is backed up by what appear to be questionable methods for the official calculation of civilian casualties, since every adult male in the blast of the weapon is often considered a combatant by default, a claim that is often disputable (Qazi and Jillani, 2012).

Under President Trump, the already suspect data for civilian casualties is much harder to obtain, given that his administration has cancelled the Obama administration requirement for reporting civilian casualties (Morgan, 2019) and greatly expanded the use of drone strikes even in countries outside of active war zones (Haltianger, 2018). So, the more precise targeting available on modern drones does not necessarily correspond to fewer civilian casualties (ibid.; Zubair Shah, 2012).

These weapons also come with a moral veneer that comes from the assumption that a more ethical conduct of war is possible using these machines. Sauer and Schörnig go on to conclude that the political goals along with the normative drives of liberal democracies necessitates that unmanned systems will continue to be armed with more powerful weaponry and that they will be given more demanding missions, which will require greater autonomy for the machine (Sauer and Schörnig, 2012, p. 370).

In addition to this, drones can also help with complex political relationships such as those between Pakistan and the United States. Drone strikes have arguably benefited the Pakistani government by providing them with a tool to attack their own political enemies while simultaneously being able to criticize the United States for those killings (Zubair Shah, 2012). In some cases, the residents in tribal areas of Pakistan are sometimes grudgingly in favour of the strikes: many favour the drone strikes over the alternatives, such as military operations or less selective bombardments by Pakistani bombers and helicopter gunships (Ibid., p. 5).

The above argument shows that we can expect the research into autonomous weapons systems to increase and these systems to proliferate into every aspect of military activity across the globe. Recent history has given ample empirical evidence to back this theory up. Even though President Obama was elected largely on an anti-war vote, it has been reported that there was an 8% increase in the use of drones during his first term and in 2011 drones were used in, '... 253 strikes – one every four days ... [And] Between 2,347 and 2,956 people are reported to have died in the attacks – most of them militants' (Woods, 2011). This trend has only increased since that time. Other countries such as Israel and Italy are known to operate reconnaissance drones. Recently, the German government has announced plans to invest in both armed and unarmed drones (Gathmann et al., 2013; Kim, 2013; Medick, 2013).

As the enthusiasm for this technology grows, a mounting opposition movement has also emerged that claims that this technology is not a cheap, easy, casualty-free means of propagating just war. Instead they claim that these technologies contribute to unjust military adventures and an indefensibly immoral push-button warfare that claims the lives of thousands of innocents caught in the crossfire. In the interest of furthering the ethical use of these technologies, it is important that we give these counter arguments our full attention.

We should note that some technologies can cause what philosophers of technology call *reverse adaptation*: '... the adjustment of human ends to match the character of the available means' (Winner, 1978, p. 229). This is where the social and political milieu of a society changes to accommodate the inflexibility of a technology, rather than waiting to deploy the technology when it is more suited to the human environment. A prime example would be the way that human societies changed due to the adoption of mass production, necessitating all manner of social upheaval that proved to be fault lines of human conflict over the last two centuries. There are many other examples of this in recent history. Think of the social disruption caused by the introduction of the automobile or cellular telephone, etc. It is obvious that autonomous weapons are again confronting our societies with the problems of reverse adaptation. These weapons are completing a trend in technological warfare begun in the nineteenth century that is making traditional notions of ethics in warfare largely untenable. These notions were always on shaky ground to begin with, but the tradition of just war that reaches back to the Middle Ages has become almost moot in its ability to influence decisions made on the modern battlefield. If this was not worrisome enough, as aging drone technologies are replaced with better, smarter equipment, the surplus will find use in civilian law enforcement duties. This will complete the circle, and technologies that liberal democracies gladly used to control their enemies will now be used in ways that challenge and potentially curtail cherished civil liberties at home. Because of this, it is vital that we engage in discussion of these technologies at every opportunity.[3]

Due to the extreme ethical impacts of these technologies, there has been an extraordinary call to ban autonomous weapons systems before they have become fully operational (Altmann, 2009; Asaro, 2008; Sharkey, 2008, 2009, 2010, 2011, and 2012; Sparrow, 2007, 2009a, 2009b, 2011). We can distil these various claims against such technologies into three main categories of proposed limitations. First, there should be limits on the authority given to decisions made solely by the machine. Second, bounds must be placed on the technological capabilities of these machines. And third, there must be restrictions placed on the deployment of autonomous weapons systems. Each one of these can be looked at from a technical, legal and/or ethical standpoint.

The question of autonomous machines deciding when to use lethal force is the most ethically challenging of the three categories of proposed limitations and as such we need to pay more attention to it than the other two categories. Asaro (2008) noted that robotic weapons systems are developing along a spectrum from non-autonomous, through semi-autonomous, to fully autonomous. As we move up the scale to full autonomy, there will be a critical step taken when we allow machines to select and engage military targets on their own with little or no input from human operators (Asaro, 2008). Doing so will cloud our ability to ascribe guilt or responsibility to anyone in the event that something goes wrong. The machine might be capable of finding and attacking targets, but it is unlikely to be capable of taking a moral stand to justify its own decisions (ibid., p. 2). So, in building this decision making into the machine, we are uploading our moral responsibility to the machine as well and *abdicating our duties as moral agents.* Furthermore, we can see that if robots are not capable of making the moral decision to kill a human being, then this situation must be avoided. In recent public talks, Asaro has begun to wonder if we ought to claim the human right not to be killed by autonomous weapon systems.

Sharkey (2010), as a robotics researcher himself, argues mainly from the view that machines are never going to be able to reliably make the right choices in difficult situations on the battlefield. Robots can barely tell the difference between a human and a trash can, which begs the question of how they are going to be able to tell the difference between an innocent civilian caught on a battlefield and an irregular soldier who is posing as a civilian. This is a challenging task for a human being and well beyond the capabilities of a machine. This limitation would make an autonomous fighting machine somewhat indiscriminate and therefore unjustifiable from a moral standpoint. In an interview, Sharkey has suggested that those funding research into autonomous weapons systems have an almost mythic faith in the ability of artificial intelligence to solve these kinds of problems in a prompt manner and that this belief is far from the reality of what these systems are capable of. 'The main problem is that these systems do not have the discriminative power to do that,' he says, 'and I don't know if they ever will' (Simonite, 2008). Again, we are mistakenly placing our trust in a machine that is actually incapable of making good moral decisions, a position that is morally suspect indeed.

Sparrow (2007) argues that it would be immoral to give machines the responsibility of choosing their own targets even if we can somehow transcend the technical problems of complex decision making and target discrimination. He asks us to consider what we would do in the case of a machine that decided on its own to commit some sort of action that, if a human had done it, then it would have constituted a war crime. In that case he argues we would find that there is no good answer when we try to decide where to affix the moral blame for the atrocity (ibid., p. 67). Asaro believes this is because, in the end, there is no way to punish a machine as they have neither life nor liberty to lose, nor would it be reasonable to assume that the responsibility for the act rested in the machine itself, or its commanding officers or even in its builders and programmers (ibid.). *Jus in bello* (the international humanitarian laws that regulate conduct of belligerents in an armed conflict) requires that there be an ability to assign responsibility for war crimes and that the perpetrators be punished. 'If it turns out that no one can properly be held responsible in the scenario described above, then AWS [autonomous weapons systems] will violate this important condition of *jus in bello*' (ibid., p. 68). Consequently, Asaro concludes that the use of this kind of weapon would be immoral and hence must be avoided.

The above arguments call into question not only the morality of having a machine decide to kill an individual human being but even their use of force to simply injure an opponent or follow opponents across political borders, as this would no doubt incite retaliation and could even lead to an escalating situation where decisions by a machine might lead to open warfare between humans. This leads the International Committee for Robot Arms Control (ICRAC) to suggest that these decisions should never be left to a machine alone.

While it is quite reasonable to seek limits on the use of autonomous weapons in situations where they could inadvertently escalate a conflict, the argument that autonomous weapons need to be banned because moral blame cannot be affixed to them is much harder to follow. Even if it were problematic to ascribe moral agency to the machine for metaethical reasons, there would still be legal recourse, and the commanding officers that deployed the weapon as well as its builders and programmers could be held liable. Of course, if these people were also shielded from culpability through various legal means, then Sparrow would be correct in making a strong claim that the lack of a responsible agent renders the use of these weapons immoral. It is not clear that that is happening yet, so this argument should be tabled until there is evidence suggesting that military commanders are claiming autonomous weapons have a rogue status that absolves anyone but the machine itself of moral responsibility.

It is difficult to find arguments in favour of giving the choice to use lethal force solely to machines. Yet it has been argued that if machines truly did have the ability to accurately distinguish between targets, then we might expect a great deal of precision in these judgments and in that case it would be moral to allow them some autonomy on the battlefield (Lin, Abney, and Bekey, 2008). Given that machines would not experience the same combat related stresses that make human judgment prone to error on the battlefield, there might be good reason to take this claim seriously. Higher reasoning powers in humans are often the first casualty when violence erupts, causing some to make regrettable decisions. A machine, for instance, would not have to instantly respond to enemy fire since it does not have a right to self-preservation. It could instead wait and fire only when it was sure of its target. Ostensibly, it would be able to deliver return fire accurately with less chance of harming innocent civilians which might marginally improve the ethical outcomes of violent encounters.

If the members of ICRAC are wrong in their assessment of the ability of these machines to discriminate between targets, then that somewhat weakens their case.

Arkin (2007, 2010) argues a more subtle point. He might agree that machines should only fire their weapons under human supervision, but he would like to see machines that have the ability to autonomously decide not to fire their weapons even when ordered to do so by humans. He would rather design a system that independently reviewed the constraints on its operation imposed by the rules of operation, laws of war, just war theory, etc., that it was operating under. This system, called an 'ethical governor', would continuously assess the situation and if the machine decided that the operation was beyond set parameters then it would disengage its ability to fire. In this way the machine's artificial ethical governor would also be able to control human decisions that might be immoral or illegal but that emotion or the heat of the battle had made the human actors unable to accurately process. In an interview Arkin said that, '[o]ne of the fundamental abilities I want to give [these systems] is to refuse an order and explain why' (Simonite, 2008). Again, since the machine has no right to self-preservation, it can legitimately disarm itself if needed.[4]

Additionally, Arkin argues that the machine can gauge the proportionality of the fire it delivers to suit the situation it is in. Where a human might need to fire to ensure that the enemy is killed and no longer a threat to his or her person, the machine can take a calculated risk of destruction and instead only would apprehend an enemy rather than always delivering lethal force. An additional strength of this design would be that it would put a safety layer on the possibility that the human operators of the machine might be making an immoral decision to fire based on emotion, stress, or improper understanding of the situation on the ground. The robot would be able to disobey the order to fire and explain exactly why it did without any fear of dishonour or court martial that a human soldier might succumb to in a similar situation (Arkin 2007, 2010; Sullins 2010a). The system Arkin proposes would be far less likely to cause the false positive errors of excessive or indiscriminate use of force that other critics worry about, but it does leave open the possibility of a false negative, where a legitimate target may get away due to situations that cause the ethical governor to engage. What if this enemy then went on to commit his or her own war crimes? Surely this would be an immoral outcome. And is most likely why we have yet to see this system deployed.

We can see, therefore, that the stance on banning autonomous targeting decisions is indeed one that requires more discussion and it is appropriate to place it on the table for potential restrictions in any AI/robotic arms control deliberations.

5. Autonomous Vehicles

Autonomous vehicles have gone from future world fantasy to headline reality at a head spinning speed. In 2005, the Stanford Artificial Intelligence Laboratory built an autonomous vehicle they called 'Stanley', which successfully completed the DARPA grand challenge. Not only did they win, but they were the first vehicle to make it through the course without crashing. This research spun off from the university into Google and by 2012 they had licensed a self-driving Prius in Nevada and by 2015, they had built their own self driving vehicle from the ground up, this one with no steering wheel at all. Today, this technology has spun off into a company called Waymo that is bringing these technologies to the mass market and plans to have autonomous taxis on the street by 2020. These successes have spurred other companies to compete with Waymo. At this point, every major automaker, and many technology companies, are racing to be the first to bring these technologies to the open road. In under fifteen years these technologies have gone from dream to almost an everyday technology.

Given that transportation is a major aspect of all of our lives, it is important to ask if these companies are paying proper attention to the ethical impacts of their work. Why might we assume that a world with autonomous cars is better than one without? The primary motivation for creating autonomous vehicles is often reported to be the claim that these vehicles will bring about a world with fewer car accidents and less of the loss of life, grave injuries, and property damage associated with them.

In 2010, the World Health Organization reported that there were 1.25 million deaths due to traffic accidents and dropping that down to zero would end a lot of needless suffering. If this is true, it would stand as evidence that we have something like a moral imperative to create safe autonomous vehicles so we can come closer to driving the loss of life associated with transportation closer to zero. But since being contradictory and contrary is part of the job description in philosophy, we need to bring up a couple of ugly counter arguments.

First, there are other equally or even more important killers in the world that are more easily solved through existing technologies. For instance, The World Health Organization also reports 1.7 billion cases of childhood diarrheal disease, which is the leading killer of children under five with approximately 525,000 of them succumbing to this disease every year. Many of these could be prevented with existing water treatment and sanitation technologies. One might say that while it is sad that children die in such a preventable way, that problem is something for others to solve and automakers must focus on what automakers can do in their own industry. This is fair enough, but it raises the question of why we put so much money into these particular technologies and whether we must always do so? This brings us to a second, truly awful point: one in five organ donations in the US come from victims of vehicular accidents, so even if we were to drop traffic accidents to zero, this would have the unintended consequence of cutting off organ donations leading to more death in other sectors of our society. This is just to say that the moral issues around autonomous vehicles are complicated and there will be no easy answers or non-conflicting moral imperatives to guide us in our deliberations.

Another troubling set of ethical questions that autonomous vehicles raise are clustered around the impacts that these technologies will have on work life and civic life. Currently there is much research to determine the extent of unemployment that autonomous vehicles will contribute to. Again, the answer is complicated, while it is very likely that we will soon see autonomous trucks platooned together safely driving at close to one hundred miles per hour with only thirty feet between them. This will be facilitated by the fact that if one of the lead trucks brakes, it will instantly cause the autonomous systems driving all those following to brake as well. So, the days of Smokey and the Bandit will be long gone as fleets of robot trucks crisscross the nation all following the laws of the road to the letter. Gone will be the long-haul truckers and the highway patrol that were needed to keep them in line. In the near future this is likely to only happen on the interstates due to the fact that human drivers will need to take over once the trucks get within urban areas. So not all truckers are going to be obsolete, just the highly paid ones. A 2018 report commissioned by the Center for Labor Research and Education at University of California, Berkeley, and Working Partnerships USA (Viscelli, 2018) concluded that 294,000 long-distance truck drivers would lose their jobs and be replaced by autonomous vehicles over the next twenty-five years. This will be a big win for trucking companies who have to deal with high driver turnover and the fact that regulations do not allow them to overwork human drivers, whereas autonomous machines could be nearly constantly on the job. Over time, as these technologies improve, other transportation jobs could be lost leading up to losses in the millions of jobs.

There has been some interest in applying ethical concepts to the design of autonomous cars. A good example of this can be found in the work of Dr. Chris Gerdes, the Director of the Revs Program at Stanford, Center for Automotive Research (CARS). Dr. Gerdes has taken a deep interest in the important role that ethics and ethics researchers can play in the design of autonomous vehicles and has had a number of his graduate students working on ways to mitigate some of the ethical impacts of autonomous vehicles. The philosopher Patrick Lin at California Polytechnic State University and his team at the Ethics + Emerging Sciences Group (http://ethics.calpoly.edu)

should be credited with gently opening the door to the CARS lab and presenting them with the value that philosophy can bring to engineering ethical systems. Since then Dr. Gerdes has gone on to contact others in the philosophical field of values-centred-design and his lab takes what they have learned and applies it to their designs.

Acknowledgement of the value of philosophical thinking in design from highly placed figures like Dr. Gerdes has helped carve a space for ethicists in the world of autonomous vehicle design. Some of the issues that come up often in the design of these systems are as follows. One of the most concerning is how to maintain meaningful human control over autonomous vehicles? We like to think that these are entities that behave like the AI car KITT in the TV show 'Knight Rider' (https://www.imdb.com/title/tt0083437), where the car has an AI consciousness programed into it and is a trusted sidekick to the hero of the show. In reality, these are (probably, perhaps) unconscious machines with less sense about themselves and the world than an ant might have. Their sense of being in the world is, perhaps, more at the bacteria level. They are simply stimulus and response machines that take actions but do not, and cannot, consciously deliberate on those actions. This level of reasoning can only be done by humans that must still play a vital role in the process, especially when it comes to taking blame when things go wrong.

For example, who is to blame for the recent fatal crash between an autonomous car and a homeless person who was jay-walking? A lot of data was released on this accident and if you view the video taken from inside the vehicle you might agree that the victim came out of the dark and it would have been difficult for even a human driver to stop in time. Patrick Lin identifies five possible locations for placing the blame or responsibility for this accident: (1) the victim, (2) the manufacturer of the autonomous system running the car, which in this case was Uber, (3) Volvo, who was the manufacturer of the car, (4) the safety driver in the car, (5) society – or no one (Lin, 2018).

In reality, placing blame in this case is a very complicated issue, and it is likely that a complex mix of all five possible locations for responsibility will be required to justly place the blame. Nevertheless, a significant part of the blame must be placed on the design and operation of the sensors that the machine is using to gain data for making choices about its operations. Even though the victim emerged out of the shadows at night, wearing black, we need to remember that the machine uses lidar, which is an active sensor that emits laser pulses that bounce off surroundings. It can be used to compute the size, shape, and distance of objects in the range of the sensors. So, shadows should not be a limiting factor as they might be for the human eye.

Even in broad daylight, one should refrain from wearing dark clothing and moving in front of a car. Since these cars rely on lidar and sense you by bouncing a burst of laser light off of you, we need to remember that dark fabric tends to absorb the laser light rather than reflecting it, making the system work harder to correctly classify you as a pedestrian. This means that, in the future, we will be entering a new world where the social mores around pedestrian and vehicle interactions are going to change in ways that the general public may not be ready for.

Most of our ethical systems in philosophy assume a human agent that has basic abilities to sense and interact with others and the world. All these assumptions are off when we are dealing with machines making decisions that impact us with life and death outcomes. Brute facts such as the one that putting on a dark clothes can make you effectively invisible to an autonomous vehicle in certain situations must be taken into account in designing social systems where autonomous vehicles and humans can interact.

Recently many discussions have occurred about whether it is ethical or not for a vehicle to prioritize the lives of its passengers over those outside the car. On one hand, it is reasonable to assume that a vehicle that is carrying you has an ethical duty to keep you safe. But on the other hand, this looks like another case where one class of people can gain more safety in their expensive autonomous cars than can those who chose to remain on bikes or on foot ... or are forced to walk by their economic realities.

A nice place to start thinking about this complex problem is with the online program Moral Machine (http://moralmachine.mit.edu/). This website explains that it was set up to be '... a platform for (1) building a crowd-sourced picture of human opinion on how machines should make decisions when faced with moral dilemmas, and (2) crowd-sourcing assembly and discussion of potential scenarios of moral consequence'. The user is presented with various scenarios that an autonomous vehicle might find itself, for example, when the autonomous vehicle has to make a no-win decision where at least some of the people or animals involved will die. In one of the scenarios we are asked whether in the case of a sudden brake failure, should the car continue ahead and hit three male athletes who are in a cross walk or veer to the next lane and hit three less fit males. Another choice might be three women and two female executives versus four homeless people and one man. In other scenarios it might be that the car itself has passengers and can veer into a wall killing the passengers rather than some set of people and/or animals on the street. In the end you are presented with results that let you know which group of affected people you valued the most and which you were most willing to sacrifice. You can even add scenarios to the test if you wish through a design feature on the site.

As time goes on and more data are collected the system hopes to develop a snapshot of what our values are, collectively, and also how a machine should behave in these tragic situations. Then the vehicles could be programmed accordingly.

This brings up the question as to whether any of the logical systems devised for ethical reasoning are complete or useful enough to be engineered into autonomous vehicles. Are systems like this capable of making decisions based on personal characteristics of various potential victims? And this raises the question of whether or not it is ethical to build a system that will make life and death choices about you based on personal characteristics. If all human life is considered equally valuable (or all animal life), then the entire premise behind the Moral Machine test is itself morally suspect, given that it is to be used to help us create machines that make specific no-win life-or-death decisions.

The reader may recognize all of the above as variations on the classic *trolley problem* thought experiment. The trolley problem, developed by Philippa Foot in 1967, asks the reader to imagine that they are a person standing by a switch on the side of a trolley track. An out-of-control trolley is moving quickly towards five people who happen to be tied up and lying on the track and are sure to be killed. However, if you pull the switch, then the trolley will be diverted to a side track and the five will be saved. Unfortunately, on that side track a single person is tied up on the track and will be sure to die. What should you do? (See Foot, 1967.)

This problem has set in motion a long-running debate in philosophy. At the turn of the last century the problem was taken up by moral psychologists who have devised numerous experiments to test reaction times and neurological processes that occur when subjects are presented with the problem. It is easy to get lost in the numerous philosophical positions that can be taken on this subject, but engineers tend to want to get to something that is more clear and can be coded into a machine.

In 2016, the creators of the website Moral Machine presented some of their results in the journal Science (Bonnefon et al., 2016). One of the interesting results was that the majority of respondents surveyed agreed that in a given situation where the car had to decide to kill its passenger rather than kill multiple bystanders that the best solution was to sacrifice the passenger. But they also said that they would not buy a car that was designed to do that since they would not want to sacrifice themselves or their children in such a scenario and would prefer that multiple bystanders were killed instead. This saddles us with the conclusion that the obvious engineering solution is to just ignore this thorny ethical issue and *let the market speak.* But it is also obvious that if we asked whether people would be willing to be on the street with cars that would sacrifice many of them to save the owner of the car, then they would object to that as well. It seems therefore that there is going to be continued debate on this topic. The question is, will any of this debate lead to a moral and just resolution?

An autonomous vehicle designer might just throw up her hands here and say something to the effect, 'Look, there is no ethical problem that an autonomous vehicle can get into that can't be solved by a good set of brakes, with redundant backups!' When confronted with the scenarios above, it just stops. That is a good plan, which we endorse, but it fails to capture the fact that there are still many more ethical issues that brakes do not solve – what if even braking fiercely, the car will nevertheless kill someone . . . no matter what?

Among the many other important, but less lethal, issues that autonomous cars raise is the issue of *data ethics.* These machines will be gathering lots of information about our comings and goings, with whom we are associating and where we are doing it and for how long. Who gets to see these data? Who owns them? How long will they be stored? Can law enforcement subpoena them? These and many other questions will keep lawyers and ethicists working for quite some time on developing practical solutions – but what of moral solutions?

Another issue is just the simple driving behaviour of the car. We all know the difference between good and bad drivers and their ability to share or not share the road with others. A good example of what this kind of programing might look like is some early tests by Gerdes and Thornton who implemented various ethical choice parameters, not for choosing an answer to a trolley style problem, but for the more common problem of driving style: should the vehicle be more or less aggressive in given traffic conditions (Gredes and Thronton, 2015)? It is clear that we are getting into questions of machine morality. These will be dealt with in more detail in Chapter 9.

Ethics may be a messy subject that systems designers would like to avoid, but as Lin argues (2015), the vehicle will have to make optimization choices during a crash and these choices all have ethical implications. We have also seen that there are many other more mundane, but no less ethical, choices that the designers and users of these systems will need to address. Autonomous vehicles are a major challenge to our ethical and moral sensibilities since they remove human agency from the driving experience. It is important that we all join the discussion around the building and use of these systems so we can create a better, more ethical, future.

6. Conclusion

The coming onslaught of AI appliances has caught the AI community and philosophy unaware. Early AI did not have to think too hard about the consequences of what they were creating; they could leave that to science fiction writers to opine on. But as the big data revolution and machine learning brought about new ways to deploy AI, ways that did not lead directly to machine consciousness, but ones that did lead to effective and useful applications. AI researchers and systems designers are finding themselves involved in more socially impactful activities beyond the world of game-playing and theorem-proving. Now their work is being used in autonomous weapons, autonomous cars, AI recommending systems, medical applications, etc., etc. Real people are living or dying based on lines of code. This has necessitated a deep, inward look at how and why we design AI and robotics systems.

9

Could Embodied AIs be Ethical Agents?

1. Introduction

The Ethics of AI is one of the most exciting areas of philosophy and AI for a number of reasons, not least of which is that we are getting to see a real-time unfolding of a process that philosophers had been predicting for decades, but which had gone largely ignored in industry until it was (nearly?) too late. The history of AI ethics is a history of warnings that seemed to be varying degrees of ridiculous until they weren't. This chapter is not going to discuss that history except as it is in service to what the future may hold. And the future as presented here is a future of possibility and potential rather than inevitability. This is an argument for where we may go wrong at least as much as where we may go right. It is an argument for why bodies matter in ethics, in AI, and in ethical AI in particular, in a way that they traditionally haven't.

2. Why Metaphysics Matters to Ethics (Broadly)

In spite of what many contemporary ethicists will tell you, ethics is intimately tied up with metaphysical and ontological claims and commitments. Take an obvious example: the contemporary debate in applied ethics about the permissibility of abortion. An enormous number of ethical evaluations rest squarely on claims about the personhood of a foetus, the moment the foetus becomes a human being, or the consciousness of a foetus. All of these are ontological claims that underlie the ethical arguments that rest atop them and could not proceed without them. Of course, ontological claims matter to ethics. This is just as true for questions in AI as it is for abortion or our duties to our environment or our obligations to others.

Many ethicists carve out a small area of analysis in which to make claims, denying that metaphysical questions are their concern. The buck gets passed. This wasn't always the case and isn't always the case now. Aristotle, Cavendish, Kant, Spinoza, and de Beauvoir are all philosophers who wrote about ethical questions and saw them irreducibly entwined with certain ontological views of the world. Yet few contemporary ethicists seem to view the work of ethics in this way, or at least they refuse to make such commitments when pressed on the matter. But it is easy to see, in the history of philosophy, various ways that the ontological commitments were necessary precursors to making ethical claims. Consider Bentham, who argued that phenomenality (suffering, generally) was a precondition to being considered as a moral agent at all. Consider John Dewey, whose *Human Conduct and Nature* was an explicit attempt to make strong ethical claims while offering a pragmatist (workaround for the) metaphysics. He argued that ethics was education: what could that mean other than that our ethical arguments are not based on a set of tidy moral laws, easy to code into a machine, but instead on lifelong processes subject to certain forces in the physical, social, and cultural world, and therefore inextricable from that world? One claim we will examine here, that ontological claims are a necessary precondition for the possibility of a robust and coherent ethics, is not controversial historically, although it may prove to be more so among contemporaries.

There is at least one other enormous elephant in the ethics room when it comes to metaphysics: free will. It is implicit in every single ethical argument that the person who is acting ethically is capable of acting ethically because they are also capable of acting unethically. My laptop is not capable of either, and neither is the bee that chased me across the quad yesterday (as far as most people would argue). It seems obviously true that ethicists should not be responsible for also arguing out the detailed nuances of free will arguments alongside our ethical analyses, especially in practical and applied ethics cases, which are often where we find AI ethics arguments. But, of course, every single argument that attempts to make practical recommendations (perhaps *especially* these arguments) are smuggling in an unexplored and assumed metaphysics around *free will*, at least! And most often, when traditional conceptions of free will are being smuggled in unexamined, so is a version of mind-body dualism that would make Descartes proud in equal measure to how it would horrify the AI-theorist accidentally subscribing to it.

3. Consciousness, Metaphysics of Mind, and Ethics

Aside from the obvious and more mundane ways that metaphysics undergirds most ethics work, there has been some work in ethics (though generally ethics that do not intersect directly with AI) that ties consciousness and mindedness to ethics in particular ways. A very brief history of some relevant work will be offered here, since we are concerned here with making some sense of a possible future in which AI systems are ethical agents.

There have been a few notable, recent attempts to tie consciousness to ethics in various ways. The May, Friedman, and Clark collection, *Mind and Morals: Essays on Ethics and Cognitive Science* (1996) offers a range of arguments that cognitive science and psychology have something important to offer moral philosophers. The authors point out that the severing of metaphysics from ethics is largely courtesy of G. E. Moore's early twentieth century claims that what people ought to do and what people actually do are too often conflated (true) and that they apparently should be studied separately (untrue). We often hear from engineers and roboticists that their job is not to do the ethics but to do the engineering (Forsythe, 2001; Sullins, 2015). In AI in particular, this generally translates to the idea that we can understand, study, and program reasoning without considering ethics. But as May, Friedman, and Clark (1991, p. 1) remind us, 'Moral reasoning is one of the most complex and difficult forms of reasoning, and any robust theory of reasoning is going to have to confront it eventually.' Not only do we pretend that engineers can build things without considering the ethical consequences of their creations (see previous chapter), but within AI we have often proceeded as if we could model and replicate human reasoning in the absence of explicit attention to moral reasoning. Cognitive science is as relevant to understanding moral reasoning as it is to understanding any other type of human cognition. Yet the intersection of work in cognitive science and work in AI ethics is a very small slice of the Venn diagram of their overlap.

A common philosophical critique of any rapprochement between cognitive science and ethics is that cognitive science will simply replace the philosophy of ethics with moral psychology or empirical approaches to studying cognition. It is worth being explicit here: this piece of the possible future history of ethical AI is not one of replacement. One must have a small view of philosophy to imagine that allowing cognitive science to say something important about ethics would erase the philosophical contributions. While an interdisciplinary approach to ethics, especially in the AI domain, seems vital, it must be repeated that this is a problem that requires the methodologies and content – the meat of the philosophy -to interact with the methodologies and content of cognitive science (including, but not limited to, other areas of philosophy, various sub-disciplines in psychology, some areas of linguistics and biology, anthropology, and engineering at least). Any argument that philosophy or ethics must be entirely replaced by cognitive science is fundamentally contrary to the argument forthcoming here. Most important to the argument here is not simply to attempt an empirical survey of how people act and attempt to draw moral conclusions, but instead to take seriously approaches to understanding human moral reasoning (and to hopefully apply this to AI imminently) that examine what Mark Johnson (1996) points out are the conceptual structures that underlie moral reasoning.

There have been few notable attempts at relating the mind to ethics in explicit ways that have made impacts in the ethics literature, probably as a result of the aforementioned boxing off of subspecialties within the field of philosophy. The interaction tends to be more or less unidirectional, insofar as more work looks at the ethical systems that arise out of various claims about the mind than starts with ethics and seeks out confirming metaphysics. Here, we will look fairly closely at various kinds of embodied metaphysics, for reasons which should become clear. Indeed, we've discussed many other approaches to AI broadly, but we've yet to really dig deeply into the implications of embodiment for various kinds of AI systems, so we'll do a lot of that work here, in concert with our explorations of ethical AI. One interesting application of work on embodied cognition and ethics is Colombetti and Torrance's (2009) work tying together a now famous piece on participatory sense-making (De Jaegher and Di Paolo, 2007) with ethics through the use of emotion as the bridge to tie reasoning and ethics together. Like many contemporary embodiment theorists (Damasio 1994, 2003; Lakoff and Johnson 1980, 1999), they don't see emotion as a bridge between reasoning and ethics, but instead they see the traditional split between reasoning and emotion to be bogus, based on poor philosophy and poor empirical psychology (cf. the discussion in Chapter 7). Colombetti and Torrance largely explore what a genuine intersubjective interactivism would mean for ethics broadly, concluding that an inter-enactive account of ethics ought to be taken as a legitimate system, due to take its place alongside the traditional views of deontology, utilitarianism, and virtue ethics. But this ethical system, rather than merely requiring lists of rules, requires certain kinds of embodied social beings (in this case, implicitly and explicitly with certain kinds of biology, but it's fair to explore how essential that is in light of materials sciences that may be able to replicate important aspects of that biology in artificial systems, although Colombetti and Torrance would likely not sign on quickly to these revisions of their work.)

One other notable attempt to acknowledge and explore the relationship between cognitive science and ethics without reducing one to the other is Mark Johnson's 1993 *Moral Imagination*. This is a text that goes to great lengths to situate itself in the history of moral theory, showing how, historically, both philosophers and people in their everyday lives have believed ethics to be a matter of determining the correct moral system of rigid laws, derived from god, or nature, or conceptions of the good applied numerically to populations, and simply applying those laws. Johnson denies this, based on considerations from cognitive science. Johnson's argument rests largely on previous work done on reasoning more generally (as opposed to strictly moral reasoning), and the ways in which our human bodies are related to how we reason (Lakoff and Johnson, 1980; 1999). We will return to this argument in greater detail in a later section (Sect. 13), but for now it is sufficient to point out how relatively rare it is for consciousness or an explicit metaphysics of mind to be explicitly called upon in establishing a theory of ethics. It does not happen commonly, but it should.

4. Ethical AI (and AI Ethics) is a Hot Mess

Much has been said elsewhere in this book about the ethics of AI applications, considered broadly (see the previous chapter). The current status of AI ethics as a field is controversial. It is also a bit of a hot mess right now, as technological challenges emerge from Silicon Valley and emerging technology groups around the world that need to be dealt with immediately (if not sooner). Tools and products and algorithms and approaches to building things all seem to hit the shelves, followed immediately by a scramble of philosophers pointing out how this or that ethicist of technology or AI has been warning us for years that this was due to happen. And consumers either never hear about these worries or continue to fail to take them seriously, compounding one ethical disaster onto another around privacy, security, or general failures to consider the social implications of our emerging technologies.

There are many questions in AI ethics that relate to military robots, sex robots, elder- and childcare robots, not to mention questions about smart appliances, ubiquitous computing, privacy and security of our online lives, as well as the adoption of some of these technologies by nation-states and militarized police forces. While there are countless entry points into the AI ethics debates right now, this chapter focuses on the relationship between the embodied/embedded/enactive/extended mind, whether human or AI, and ethical agency. Here, the exploration and argument are about the ethics *of the machine*; the capacity the machine itself has for possibly performing ethically, rather than our ethical duties to it or the implications of its existence in our ethical sphere. Here the future of ethical AI has something really interesting to offer us, as it promises to tie together the literature in '4e' cognitive studies (the four 'e's being embodied, embedded, enactive, and extended) with the literature in robotics, and to offer us insight into the nature of ethics itself. Importantly, there is one major assumption running throughout this chapter, and that is that it is only worth considering the possibility that a machine could have ethical agency in the way we understand humans to have it if that machine has something like what we would call a mind. The word 'consciousness' may be used interchangeably with 'minded' here, but certainly no one has solved the deeper question of consciousness (see Chapter 7). A view will be examined and at least somewhat defended that consciousness is immediately and inextricably tied to embodiment, though, so this breaks with most traditional computationalist views and remains highly controversial (even to some of the authors of this text).

5. Traditional AI Ethics: Hard Coding Ethical Rules

Given the perspective being taken up here, that we must pay some attention to how a system works and what the underlying possibilities of that system are in understanding the possible ethics of that system, it is worth taking some time looking at how the 'war' about ethical AI has unfolded up to this point.

Think back to Asimov's Three Laws of Robotics. They are patently absurd, but they give a good sense of where traditional ethical AI started and where a good chunk of the field remains. Asimov's laws are fairly straightforward: 1) No machine may harm a human or, through inaction, allow harm to come to a human. 2) A robot must obey orders from a human except insofar as those orders conflict with the first law. 3) A robot must protect its own existence as long as that protection does not interfere with the first or second laws (Asimov, 1950). The most notable fact about these laws is that they were understood to be something that could be *hard coded* explicitly into the systems as they were produced, so that every robot in Asimov's universe came into the world with these laws as a default part of the code. For a long time, this was the basic approach to ethical AI in the real world, for a number of reasons. First, consider that GOFAI (Good Old Fashioned AI, aka, *symbol processing*, see

Chapter 4) was the only game in town for a while. The idea of hard coding the laws of ethics into a machine was more or less uncontroversial, because for a while the pursuit of AI simply was the idea of hard coding laws into a machine. Which isn't to say there were no red flags raised historically at this approach, but that for a while, there simply wasn't much to do otherwise.

The year 2009 saw the publication of what is now one of the contemporary classics of ethical AI scholarship: Wallach and Allen's *Moral Machines*. Between the GOFAI days of the 1960s and 1970s and 2009, you can see that not too much had changed with regard to the general approach of ethical AI, in that talk of hard-coding ethical systems into a machine was still an option under serious consideration. On the other hand, you can see throughout that text how contemporary philosophers have grappled with that legacy. Importantly, of course, Wallach and Allen hang on to *functionalism* (mental states such as belief or desire are the states that they are due to their relation to other states) and its assumptions as a vital part of the story they tell, which moves us through how they predict artificial moral agents (AMAs) will develop. It is an important history, and it is almost certainly not entirely wrong. They imagine artificial agents moving through operational morality, into functional morality, before achieving full moral agency (p. 26). And while they do gesture toward alternative approaches to artificial ethics (including their Chapter 10, where they entertain questions of embodiment, consciousness, and empathy), ultimately it seems they still endorse a largely traditional computationalist view of artificial ethics. (Computationalism is the only surviving branch of functionalism; see Chapter 4.)

This has remained a popular approach, and hard-coding ethical rules into machines (at various levels) retains a strong metaphysical commitment to several ideas, namely that we humans are, at a fundamental level, computers, and that includes our ethical reasoning systems. We can look briefly here at several examples of these kinds of views.

An Example of Traditional AI Ethics at Work

A recent (2015) piece by Govindarajulu and Bringsjord argues that ethical control must be built into the system at the operating-system level. They are speaking specifically here about robots, but considering the system they've imagined, the body does not play a significant role in the ethical decision-making they envision. This is important and is something we will return to below: while many arguments are made now about robotic systems rather than systems of pure code, the underlying assumptions are almost always such that the robots are incidental structures that play no real role in the content or decision-making processes of the systems. The fact of the body in these sorts of robot-ethics cases generally serves only to change the optics of the case by appealing to our anthropomorphizing of the robots. Govindarajulu and Bringsjord claim, 'Ideally, the ethical substrate should not only vet plans and actions but should also certify that any change (adding or deleting modules, updating modules, etc.) to the robotic substrate does not violate a certain set of minimal ethical conditions' (p. 88). This reference to a 'set of minimal ethical conditions' smuggles in the assumption (or in this case, really highlights it) that such ethical rules exist and can be built in (either on the operating system level or even in basic rules of action embedded in the higher-level code). The spirit of Asimov's Laws is alive and well here.

While it is occasionally acknowledged that traditional moral theories (deontology, virtue ethics, and utilitarianism) have limits with regard to the messiness of real human moral agency, most AI approaches continue to cling to a computational structure as the architectural limit that must be respected in the quest for a workable, functional moral approach to AI ethics. For example, Bello and Bringsjord (2013) make a detailed plea to AI ethicists that we pay attention to moral psychology and cognitive architecture. They (correctly) acknowledge the messiness of traditional approaches but lump together the computational-friendly calculus of utilitarianism and deontology with virtue ethics, eliding the ways that virtue ethics is not a simple calculation and may not be tractable in the same ways. It is clearly true that moral psychology has much to teach us about how humans act and why we act in those ways, and hence it is the right source of data about the possible architecture of our moral reasoning systems. However, any plea to the current empirical data in cognitive psychology must also begin with an acknowledgement that the information-processing model of human cognition remains the dominant paradigm in cognitive psychology even as it comes under scrutiny in every other psychological domain. In other words, even as they responsibly try to draw from data in experimental psychology, most AI ethicists fail to acknowledge the (potentially) dysfunctional metaphysics that undergird the entire enterprise: strict, old-fashioned computationalism.

6. Consciousness as Ethical Grounding

So far, we have seen that metaphysics has historically been traditionally tied up with various ethical systems, although most contemporary ethicists attempt to avoid these (unavoidable) entanglements. Further, we have seen that the metaphysics of mind in particular has been occasionally tied to some ethical systems and analysis, and indeed historically this was the norm (Bentham, 1970; May, Friedman, and Clark, 1996; Johnson, 1993). But an important question remains: to what extent has the metaphysical status of consciousness or mindedness informed or been implicated in ethical AI, in particular? A surprisingly small amount. To be fair, AI ethics covers a lot of ground, but it is surprising that more theorists aren't considering that consciousness itself may be a grounding for the possibility of ethical agency (cf. the discussion of consciousness grounding 'mattering' in Chapter 5). Here, we'll lay out briefly those who do explicitly or implicitly make such claims and examine the role that the metaphysics of mind may play in the future of ethical AI.

As one of the few writers to tackle this question, Torrance (2008; 2014) describes the landscape even as it continues to evolve. He introduces (although does not necessarily endorse) what he calls 'the organic view', which is a claim based on the (often unclear) notion of sentience: that no computational system could be sentient, and therefore would never be capable of displaying 'empathic rationality', which may be deemed necessary for being considered a moral agent. Torrance suggests that moral patiency may be a prerequisite for moral agency, and that full moral agency requires this 'empathic rationality', which he contrasts with a more intellectual kind of rationality that most of us have no trouble attributing to computational systems even now. This empathic rationality is one that 'involves the capacity for a kind of affective or empathic identification with the experiential states of others . . .' (Torrance, 2008, p. 510). This argument seems to rest in part on something akin to Thompson's (2001) linking of empathy and consciousness, where he argues that empathy is a precondition of the very possibility of consciousness. These arguments turn up more often in the literature labelled 'enactivist' rather than 'embodied', but there are useful overlaps here (see Chapter 4). It is worth pointing out that there is likely a way to take inspiration from, or to buy some of these claims, without necessarily committing to the full suite of implications of this work. (In other words, do we really require these squishy meat bodies and precise neural systems to see some of these principles through? Or could robotic systems replicate the important principles if we could just settle on what those are, even though historically we have bungled it time and again?)

In other places (Torrance, 2014), the argument is more explicit in drawing out the organic view as that of being a realist about the metaphysical nature of experiences, which are either absent or present, in making possible the status of moral patiency that is the precondition for moral agency. Important to note here is the distinction, that follows us all across the history of AI, between the attributed view and the realist view (in this context, what Torrance calls the realist versus the social relationist views). It is the problem we see in every aspect of the strong AI/AGI project: that we may attribute certain states to the system based on behaviours it exhibits and our knowledge of the underlying workings of the system, but all of the interesting stuff happens when we query as realists: does the system actually have those states, or just appear to? In this case, at least some have argued that we will continue to attribute those states to increasingly complex AI systems, but perhaps any appearance of ethics (and/or experience and/or consciousness itself) may not be actually present in the system. This is not a worry that goes away until we draw a metaphysical line in the sand: in what does ethics (in this case) or consciousness (in the most interesting case) consist such that we believe we have it and we can hope to put it in a machine?

It has similarly been argued that moral reasoning is merely another kind of reasoning, and hence any system to which we might ascribe rationality and access to the kinds of information that form moral considerations would by default count as a moral agent. So if the realist argument, that consciousness is a prerequisite for genuine moral agency (or that there even is such a thing as genuine moral agency out there in the world to be discovered), strikes you as dissatisfying, there are alternatives that rest more squarely on the side of attribution-as-realism. One such argument (Chopra, 2011) takes up Dennett's Intentional Stance and draws the logical conclusion that there also exists a moral stance we might take toward a system, and this requires us simply to ascribe to it moral beliefs and desires, making the system genuinely moral insofar as our ascription of that stance is useful to predicting and making sense of its behaviours. This seems like a clever move, but one that will ultimately be subject to all of the same critiques Dennett's own view has seen over the years. It too, however, remains a metaphysical claim about the nature of a system that might be capable of genuine moral reasoning (even if it leaves many of us dissatisfied with what we call 'genuine').

7. When is a Robot a Moral Agent?

John Sullins (2006) takes a pragmatic approach to the question of consciousness and free will as it applies to artificial moral agency. This starts with the realization that when AI comes into our lived space in the form of robotics, it can be designed in such a way that it is easily anthropomorphized. For example, a robotic system designed to take over aspects of caregiving to elderly patients is going to need some sort of programing to fill in for the ethical choices that a human caregiver would have to make if they were doing these tasks instead of the machine. When this happens, he argues, there are three initial answers to the question of whether or not the machine can be said to be a moral agent:

> The first possibility is that the morality of the situation is just an illusion. We fallaciously ascribe moral rights and responsibilities to the machine due to an error in judgment based merely on the humanoid appearance or clever programming of the robot. The second option is that the situation is pseudo-moral. That is, it is partially moral, but the robotic agents involved lack something that would make them fully moral agents. And finally, even though these situations may be novel, they are nonetheless real moral situations that must be taken seriously.
>
> Sullins, 2006

The first option, that artificial moral agents are an illusion, is one that seems the most rational. If AI and robots are just computational tools, then it follows that any perceived ethical or unethical action produced by the machine is at best a reflection of the ethical desires of its programmers. While this looks good on the surface, Sullins argues that it is not as simple as that given that technologies, even unintelligent technologies, do colour the ethical decisions of their users. While it is widely repeated that 'guns don't kill people, people kill people', it must be acknowledged that guns make it a lot easier to kill someone and suggest the possibility of killing others by their very design. When we add lethal autonomous weapons to the discussion, it is not clear at all that the weapon itself might not have a significant role to play in who does or does not get killed. We can see that technologies are not a simple conduit of ethical desire, but that they can also amplify, mute, alter, and even confound the ethical desires of their designers and users.

Sullins argues that what we want from AI and robotic technologies is more similar to what we get from domesticated animals than it is to simple technologies. His example is that we will have to assess the ethics of an AI system in a similar way to a *technology* like the guide dog.

> Certainly, providing guide dogs for the visually impaired is morally praiseworthy, but is a good guide dog morally praiseworthy in itself? I think so. There are two sensible ways to believe this. The least controversial is to

consider things that perform their function well have a moral value equal to the moral value of the actions they facilitate. A more contentious claim is the argument that animals have their own wants, desires and states of wellbeing, and this autonomy, though not as robust as that of humans, is nonetheless advanced enough to give the dog a claim for both moral rights and possibly some meagre moral responsibilities as well.

In the guide dog situation, we have trainers and breeders, who interact with the dogs, who then interact with the visually impaired person to be helped. If we imagined some near future technology that played the role of a guide dog, we would have designers, engineers, and programmers who create and program the machine that then interacts with the visually impaired person. Any pro- or antisocial outcomes would need to be ethically addressed in a similar way. This line of thought is motivated by the philosopher of technology Carl Mitcham who states that the '... ontology of artefacts ultimately may not be able to be divorced from the philosophy of nature' (Mitcham, 1994, p. 174). This might lead us in the direction of the third possibility enumerated above, in which we must admit that some (but not all) actions that are initiated by AI systems can be considered ethical acts.

Sullins lists four positions held by various philosophers who have taken stands on this subject. Daniel Dennett (1998) made a short statement where he concluded that an AI machine could be built today that had the capacity for committing immoral acts, such as murder, since it could be designed to have *mens rea* or a guilty state of mind where its motivational states, systematic belief sets, and non-mental states of negligence, could lead it to commit wrongful murder, but that this system would still lack 'higher order intentionality', which is required for the AI agent to be culpable for murder. Culpability would be left to the reckless designers and users of such a dangerous system. Dennett does not rule out that higher order intentionality is possible in future machines only that we are not in any position to be able to give it to current technologies.

A second position can be held that comes to a similar conclusion but rests on the idea that AI will never have an autonomous free will, which is essential for moral agency, thus no AI will be a true moral agent since its actions are programmed or at best random, but never a result of their own consciously made choices amongst well understood options. Sullins uses the work of Selmer Bringsjord (2007), as an illustration of this view. It is worth noting that later work by Bringsjord does allow for what we might call *artificial ethical agents*, or systems that help us achieve ethical (or unethical) actions but can only do so with the help of proper (or improper) programming.

Sullins cites the work of Bernhard Irrgang (2006), as a representative of a very similar position that adds the twist that it admits the possibility of a cyborg (AI enhanced human) as a location for AI moral agency but that a standard AI could not be a moral agent given that it has no subjectivity or personhood.

A third, and very interesting option, is that Earth has never actually seen a moral agent, given that humans are incapable of being one, but that it soon will when AI finally evolves to the point of becoming the first moral agents on this planet. This strangely wonderful argument was put forward by Joseph Emile Nadeau (2006). He agrees that only a free agent can make truly moral choices, but his criteria for true freedom is that the action must be the result of *strict theorem proving* in a rigorously logical way. Human 'moral' reasoning is based on emotions and hormones, chance, and sentiment, so none of us has ever committed an act of free will that was not determined by these evolved psychological mechanisms. AI, however, would be free from all that and could use *pure reason* to determine its actions based on accurately weighed data. The actions of such an agent would therefore be *bona fide* moral acts that were fully justified and verified by logic.

We do not have such machines, but Nadeau saw no reason why we could not eventually build one. Sullins also reminds us of the position taken by Luciano Floridi and J. W. Sanders (2004) who argued that we need to find a way around the many philosophical paradoxes and rabbit holes surrounding concepts like 'free will', 'consciousness', and 'intentionality' by accepting that these concepts are inappropriately applied to artificial agents such as AI and that some seemingly mindless agents can, under certain analytical circumstances, be seen as actual moral agents.

Floridi and Sanders ask us to consider how levels of abstraction or levels of analysis change our conception of agency, even in humans. When we focus our analysis on a human being and get to the point of seeing only biological systems and other mechanical operations of the body, we lose all sense of the agency of that being and we certainly lose sight of the moral capacities that that person might have. It is only when we expand our analysis back out that the agency begins to come into focus. The concept here is that anything can be abstracted to the point that it *only looks like* a mechanical system, but some entities can be seen as agents when the systems are analysed at different levels (cf. the 'systems reply' to Searle in Supplement 5.2). If an entity is interactive and adaptive to its environment at some levels of abstraction, then that entity can be considered an agent, even if at other levels of abstraction it does not have these qualities. Agency is something like an emergent quality that expresses itself under the correct systematic and environmental conditions.

> When these autonomous interactions pass a threshold of tolerance and cause harm we can logically ascribe a negative moral value to them, likewise the agents can hold a certain appropriate level of moral consideration themselves, in much the same way that one may argue for the moral status of animals, environments, or even legal entities such as corporations.
>
> Floridi and Sanders, paraphrased in Sullins, 2006

From this analysis of allied philosophical arguments for artificial moral agency, Sullins then adds his contributions by providing a set of conditions that, if met at an appropriate level of abstraction, suggest that the agent in question is a moral agent, or at least an agent that has the capacity to make certain ethical decisions. The criteria that must be met at the appropriate level of abstraction to determine moral/ethical agency are:

1 Is the robot significantly autonomous?
2 Is the robot's behaviour intentional?
3 Is the robot in a position of responsibility?

The *autonomy* criterion can be met by a system that is not under the direct control of a human operator during the time of its making decisions that have ethical impact. There is no need for deep philosophical autonomy here; this is not a call for radical free will or uncaused actions. Instead it is just the simple question, is the system acting entirely under its own programming? This criterion does allow for the complexity of AI/robotic systems, where a robot may be controlled at a distance by an AI system, and a system may be composed of numerous individual components. All that it requires is that the system, however simple or complex, can be seen as acting autonomously from the control of some human agent. So your standard car is out, your autonomous car that passes control back to you when things get difficult are not, and an autonomous vehicle that maintains control and makes decisions in emergency situations while you are just along for the ride would definitely show the proper level of autonomy.

The criterion for *intentionality* (in the ordinary sense, not the 'aboutness' of Chapter 5) also tries to dodge some of the more complex philosophical interpretations of what makes an act intentional. Certainly, many standard ethical theories tend to stress that ethical actions must be intentional. That is, the action must be one that the agent knew it was doing and had a reasonable prediction of the eventual outcome of that action. Sullins worries that given that we have a hard time ascribing intentionality to human agents without argument, for instance hard determinists could not accept intentionality without a great deal of philosophical gymnastics, so it is best to stick with a functionalist account of intentionality and as such he can claim:

> If the complex interaction of the robot's programming and environment causes the machine to act in a way that is morally harmful or beneficial, and the actions are seemingly deliberate and calculated, then the machine is a moral agent.

> Sullins, 2006

This means that if the AI/robotic agent displays a level of intentionality, and seeming free will, that is roughly equivalent with what human agents display in those same situations, then actions that would be considered intentionally, ethically motivated if done by a human must also be accepted as intentionally, ethically motivated in an AI/robotic system.

The final criterion is *responsibility*, meaning that the social situation or milieu that the AI/robotic system is interacting in is one that includes certain ethical responsibilities.

> If the robot behaves in this way and it fulfills some social role that carries with it some assumed responsibilities, and only way we can make sense of its behaviour is to ascribe to it the 'belief' that it has the duty to care for its patients, then we can ascribe to this machine the status of a moral agent.

> Sullins, 2006

This is the recognition that the way we use an AI/robot system has important moral significance. If we are handing over sensitive duties to these systems and those duties come with socially understood ethical responsibilities, then we are at least treating these systems as if they were moral agents.

It is very important to remember that Sullins is not necessarily endorsing that we indeed build systems that have autonomy, the intention to make ethical decisions, and are in a position to do so with consequences that cannot be reversed. Not every moral agent is a good or effective moral agent. The argument here is that it is just not impossible for us to do this. The question of whether it is prudent or not is separate. That is the question of whether we want to cede our ethical responsibilities to machines. This question is in hot debate, and this chapter is meant to help clarify it.

8. The Role of Consciousness in Artificial Ethical Agents

One of the strongest claims that machines lack something that biological agents have and therefore cannot be moral agents comes in the form of a lack of consciousness. Previous chapters have considered the various philosophical arguments about the possibility or impossibility of creating machine consciousness. It is fair to suggest that this potential lack will make it impossible for artificial agents to attain the status of full *Artificial Moral Agents* (AMA). It is not so clear that this argument holds when we consider other categories of artificial agents in relation to their ethical status such as *ethical impact agents* (EIA) and *artificial ethical agents* (AEA), but in this section we will refer only to the most advanced category, the AMA (see Sect. 15 below for more details of this classification). Steve Torrance addresses this concern in his research (2008). He thinks this is an important question to discuss, as:

> This is the question of artificial agents as potential bearers, in themselves, of inherent ethical duties and/or rights – that is, as agents which (or who?), on the one hand, (1) are considered as themselves having moral obligations to act in certain ways towards us and each other; and who, on the other, (2) might themselves be fit targets of moral concern by us (that is, as beings with their own moral interests or rights, as well as moral responsibilities).

Torrance is primarily concerned with the many ethical problems that would come with the creation of such artificial beings. It would require the expansion of our moral universe to include these entities and we have only just recently, with the International Declaration of Human Rights (1948), gotten around to expanding the moral universe to include all humans and we are not in any kind of agreement on the moral status of animals or the environment. So, the addition of artificial agents is likely to stress our already not well-developed moral intuitions.

One of the main intuitions that Torrance believes will play a role is the common sense belief that shared consciousness is vital to humans accepting each other as moral equals.

> ... it might be said that my ethical attitude towards another human is strongly conditioned by my sense of that human's consciousness: that I would not be so likely to feel moral concern for a person who behaved as if in great distress (for example) if I came to believe that the individual had no capacity for

consciously feeling distress, who was simply exhibiting the 'outward' behavioural signs of distress without the 'inner' sentient states. Similarly it might be said that I would be less prone to judge you as morally responsible, as having moral obligations, if I came to believe that you were a non-sentient 'zombie'. So there may be a kind of internal relationship between having moral status and having qualitative conscious states.

Torrance, 2008

If Torrance is correct, and given the current state of our (in)ability to even describe what consciousness is and is not, and our inability to definitively state which entities in our universe have consciousness and which don't, plus the fact that we can't measure how much of it anyone has even in fellow humans, it is highly unlikely we could convince sceptics that a machine was conscious, even if it behaved with exquisite accuracy as if it were conscious. The paucity of the evidence for believing in machine consciousness leads Torrance to jettison the claim that consciousness is necessary for AMAs and he instead seeks to argue for the moral status of machines without any reference to consciousness. We will look at how he addresses this new problem in the next section.

9. The Ethical Status of Artificial Agents – With and Without Consciousness

If we agree with Torrance that consciousness is not going to help us decide the moral status of artificial agents, then we are left with the problem of philosophically justifying the moral agency of non-conscious entities. Torrance (2008), presents what he calls the 'organic view', which makes these five claims:

1 There is a crucial dichotomy between beings that possess organic or biological characteristics, on the one hand, and 'mere' machines on the other.

2 It is appropriate to consider only a genuine organism (whether human or animal; whether naturally occurring or artificially synthesized) as being a candidate for intrinsic moral status – so that nothing that is clearly on the machine side of the machine-organism divide can coherently be considered as having any intrinsic moral status.

3 Moral thinking, feeling and action arises organically out of the biological history of the human species and perhaps many more primitive species, which may have certain forms of moral status, at least in prototypical or embryonic form.

4 Only beings that are capable of sentient feeling or phenomenal awareness could be genuine subjects of either moral concern or moral appraisal.

5 Only biological organisms have the ability to be genuinely sentient or conscious.

In many ways this stance grows out of the criticisms of AI as expressed by John Searle (see Chapter 5): there is something possessed by biological brains that just cannot be captured through computational means. Torrance does not necessarily assert that the organic view is correct, and he acknowledges that it might be found to be wanting in future debates, but until it does, he thinks it is a strong argument against granting full moral status to artificial agents.

Torrance (2011) extends his exploration of artificial moral agency to four additional perspectives beyond the organic view. These include: anthropocentrism, infocentrism, biocentrism, and ecocentrism. Each of these philosophical stances can colour one's ability to accept artificial agents as potential moral agents. These concepts are imported from environmental ethics and may have some bearing on the issue of artificial moral agents given that environmental ethicists also have to come to terms with the moral status of environments, which are also complex systems that seem to hold value but lack consciousness in any robust sense.

In terms of how they apply to the debate about artificial moral agency, Torrance claims that *anthropocentrism*, the assigning of value to the world only in so much as that value translates directly to human well-being, would demand that machine ethical systems only have worth in that they are found useful by humans. In the absence of that value, no machine would have a claim to moral rights or privileges. Under this view it might not be wrong to build machines that operated in ethical domains, but it would be wrong to mislabel them as AMAs, no matter how useful they might or might not be.

Infocentrism places value on the information complexity of a system. Thus significantly complex systems might be seen to have their own distinct moral worth. Adherents to this view, Torrance argues, would be amenable to accepting certain complex machine entities as moral patients and possible moral agents as well. It might be imperative, with this view, that we make sure that our artificial agents have the capacity to reason ethically, lest we unleash powerful machines on the world with no regard for the lives of others.

Biocentrism places prime moral value on biological creatures. It argues that human moral sentiments did not drop to us from heaven, but instead evolved over long periods of time on this planet and therefore our morality is continuous with that of other biological creatures we have shared and now share the world with. This stance would be sceptical of machine ethical agency unless that agency could be proven to be equivalent to biological systems in almost every way. Since that is not the case with modern computer technology, the biocentrist would not accept an AMA.

Ecocentrism is a strong counterargument to how ethics has been conceived of so far. It places primary value on large earth-spanning systems such as Lovelock's Gaia Hypothesis (Lovelock, 1979). Under this ethical stance, AMAs would seem to be a big distraction from the vital work of protecting the planet and unless they could be shown to be an integral part to solving the destruction that humans have created, then they would not be desirable.

10. The Better Robots of our Nature

Eric Dietrich (2011) takes on the challenge of ecocentrism as discussed in the last section. He argues that humans have already proven beyond a doubt that they are not moral enough to be proper stewards of this planet. In the same way that we are epistemically bounded (limited in what we can know), it follows that we are morally bounded as well (limited in the moral sentiments we can conceive of and experience, hence limited in how good we can be). It is these two limits that cause us to be extremely harmful to the planet, which is of ultimate ethical worth. Therefore, we need to replace ourselves with less epistemically and morally bounded creatures.

> Put another way, the moral environment of modern Earth wrought by humans together with what current science tells us of morality, human psychology, human biology, and intelligent machines morally requires us to build our own replacements and then exit stage left. This claim might seem outrageous, but in fact it is a conclusion born of good, old fashioned rationality.
>
> Dietrich, 2011

This view is the idea that we are morally required to preserve what is good about humans, but in addition we are morally required to eliminate what is ethically reprehensible about ourselves. We can best achieve this by rigorously pursuing the design and development of AMAs. These machines would be able to discover truths about the moral landscape that we can't see in the same way that we can know so much more about calculus than the typical house pet. What we have learned through our intelligence is worth preserving, but our biological limits and human psychology are not, so we need to find a way to preserve the one and eliminate the other. The best way to do this is to build machines that can preserve and expand our intelligence and once we accomplish this, it will be time for us to get out of their way, perhaps though mass suicide or just a simple die off. After this, the planet will be under better hands and will be able to survive as a deep green ecosystem for the maximum amount of time possible (until the sun expands and destroys the inner solar system and perhaps beyond – about 4 billion years from now; however, our sun is slowly increasing in intensity; astronomers think that in a mere 1 billion years, our sun will be so hot that Earth's oceans will boil.)

11. Bodies – Finally

As we've seen, bodies are rarely mentioned in scholarship on ethical AI in particular, and only occasionally discussed in relation to the ontologies that underlie various ethical views and arguments at all. There is one domain of AI Ethics where bodies do commonly appear as a vital part of the conversation: social robotics. This is true of attempts to build robots that emulate (or perform) real social interactions with other humans, as well as robots built in order to study social engagement and interactions among other humans. Yet, the ways that bodies routinely turn up in this literature are generally problematic, much the same way bodies in the all-human/non-robot world are: racism, sexism, and other distressing features carry over easily into our interactions with robots.

It has been noted (Shaw-Garlock, 2014) that people readily apply their own gender stereotypes when interacting with social robots, which are largely designed to draw on and engage those stereotypes. Robot bodies are often designed with these ideas in mind, drawing on hair-length, mouth-shape, voices, and even using context to make bodies matter differently in different environments (ibid; Eyssel and Hegel, 2012; Carpenter et al., 2009; Otterbacher and Talias, 2017). Much work in AI ethics has been devoted, in recent years, to noticing the social implications of our interactions with autonomous technologies such as social robots and anthropomorphized algorithms designed also to draw out our stereotyped engagement. See, for example, the ways that programs like Siri and simple GPS mapping programs use female names and voices to various, generally stereotyped effects (Eyssel et al., 2012; Nass and Brave, 2006; Siegel, Brazeal, and Norton, 2009).

Indeed, arguments have been made that embodied robots will be necessarily treated differently from other artefacts, and therefore the forms that robots take will and do impose a particular kind of ethical demand on us. Scheutz and Crowell (2007) have argued that particular embodied forms raise questions (including ethical questions about 'design, deployment, and use' (p. 1)) that ought to be explored before the robots are designed, deployed, and used in ways that might be unethical. (For example, can embodied robots ever ethically give the appearance of having a human-like mind when we know, from the design side, that no such mind is present?) This is yet another way that embodiment crops up in AI ethics literature, and it continues to inform questions about humanoid shape in design alongside questions about gender. Hanson Robotics' humanoid robot, Sophia, has been the subject of much debate within the AI community and, importantly, on the public stage, as the robot makes public appearances with pre-scripted lines, designed to give the appearance of true intelligence. Sophia was famously awarded Saudi citizenship in 2017, in a move that put gender considerations on the front page of the AI debate, considering women in Saudi Arabia still face heavy restrictions with regard to their own autonomy (Sharkey, 2018; Wootson, 2017).

In addition to gender cues, there is some evidence that people perceive the colour of humanoid or anthropomorphized robots as racialized, with all of the cultural biases that racialized skin in the US (and likely, the West more broadly) already carries with it. For example, one study used a test similar to the famous implicit bias task to show that people react differently to robots perceived as black versus white, especially when those robots are seen holding weapons (Bartneck et al. 2018). This bias can be seen in other contexts and cultures as well, as, for example, when German subjects show less preference for a robot that is introduced as Turkish (Eyssel and Kuchenbrandt, 2012). Importantly, we must recognize that there are ethical questions surrounding the form of embodiment that we design robots to have that can and must be explored in greater detail before we continue designing them. Most robots that are manufactured for commercial and consumer use are white (Reik and Howard, 2014). When Americans automatically read this as racially white, or, conversely, read black robots as racially black, there are important ethical considerations that designers are failing to engage with.

So, we can see that bodies are considered in some practical (and particularly social) AI ethics research, and indeed these examples offer evidence that we must consider them further, particularly when we are designing social robots that will interact with humans. But there are other aspects of AI ethics work that fail to consider the impact and importance of embodiment generally. In the next section, we'll talk about why more work in AI ethics ought to consider cognitive science literature on embodiment more generally.

12. Embodiment

There is a great deal of debate in both philosophy and psychology over what the term 'embodiment' means. It is deployed widely in often-conflicting ways, and trying to comprehensively taxonomize or tease apart these uses into distinct research programs is well-beyond the scope of this book. However, we can at least get clear on the general way the term is used here and offer a few historical looks at how the term has been shaped. It is worth pointing out that even the term 'embodied' is somewhat controversial: the concept overlaps with a number of similar views, and all are related to the way we use 'embodiment' or 'embodied mind' here. See also the terms: situated cognition, situated action, embedded cognition, embodied cognition, radical embodied cognitive science, extended mind, enactivism, and distributed cognition. (See Chapter 4.)

Probably the most useful taxonomy of the term embodiment offered in psychology comes from Wilson (2002), who offers six distinct ways the term is used in the literature.

First, cognition is situated. This simply means that cognition occurs in a context where new inputs and outputs are being generated and enacted in real-time, in relation to whatever task the system is currently engaged with.

Second, cognition is time-pressured. This means that the cognitive system cannot take all the time in the world to build up representations of the environment, or it will fail to execute its tasks in a reasonable manner. Embodied robotics tends to be a great example here, as imagining a robot walking on uneven ground, we can quickly see why time-constraints matter so much. This version of embodiment seems to echo earlier AI concerns about the frame problem (see Chapter 6).

The third way 'embodiment' is deployed is as a way to describe how we offload cognitive work onto the environment. This view is similar to Brooks's

(1991) famous claim that 'the world is its own best model'. This happens when we leave information out in the world instead of trying to represent all of it in some sort of internal model. Many humans count on their fingers or rearrange Scrabble tiles rather than attempt to shuffle that information purely through mental representation. Repeated experiments in psychology show how *change blindness* and *inattentional blindness* occur when we're surprised by information we failed to encode changing in the real world in unexpected ways (this is a version of the frame problem, see Chapter 6). The extent to which we do this, and for what kinds of information, is still unclear.

The fourth version of embodiment, according to Wilson, is when the environment is actually a part of the cognitive system. This version is the one that tends to be called 'distributed cognition' in some circles now, though of course that term is just as contested as 'embodiment'. This is also likely where the contemporary view of 'extended mind' would fit in this classification, although again, none of these terms is uncontroversial.

Fifth, rather than just describing the way cognition is situated, one can emphasize the way cognition is necessarily *of* the body, specifically *for* action. By this interpretation, for example, vision is not a system used to build up an internal representation of the world, but instead used to guide action and prime motor activity.

Finally, Wilson's taxonomy includes a sixth way the term is used, generally to refer to research that catalogues the ways that offline cognition is body-based. This sixth use of the term refers to the wide body of psychological research that includes both neural simulation studies (Gallese and Sinigaglia, 2011; Gallese and Lakoff, 2005) and conceptual metaphor work (Lakoff and Johnson, 1980; 1999).

Another useful way to think about embodiment comes from Wilson and Golonka (2013), where they push back on the ways the term 'embodiment' has been used and try to narrow down a useful framework for thinking about 'embodied' as a descriptor for research going forward. They offer four key questions that we should ask ourselves to frame our thinking about this problem, starting with trying to determine what task needs to be solved, since embodied solutions solve specific tasks, not general problems. If we aren't talking about a specific task, they claim we can't usefully use the term 'embodiment' in research, and this seems like a useful way to get a much more specific frame for using the term. They also challenge us to determine what specific resources the organism has access to in order to solve the task, as embodiment draws on different resources than other kinds of tasks. They then ask us to consider how we can assemble the available resources into a type of dynamic system that can actually solve the problem, and finally, ask whether the organism does, in fact, assemble and use the resources in this way. While their view is not as definitional as the previous taxonomy, using this as a way to think about embodiment is a practical, task-based assessment that ought to help us consider what work the term is even supposed to be doing when we use it to describe research.

Again, these are only two of the recent attempts to taxonomize the many ways the term 'embodiment' is used, and none of them quite captures the way it is being used here. Shapiro (2011) collapses some of these approaches into three main camps of 'embodiment', using classical texts in cognitive science as touchstones to certain usages of the word, starting with Varela, Thompson and Rosch (1991), who use embodiment to challenge representational views by claiming that cognition is tightly dependent upon certain kinds of bodies with certain perception and action capabilities, embedded in socio-cultural contexts. This is the view that most often is attached to the term 'enactivism' now, but still often just gets called 'embodiment'. Shapiro then offers Thelen et al.'s (2001) dynamical systems approach as distinct here, calling it 'representation lite' (Shapiro, p. 56). He describes this view as involving tightly coupled processes between body, perception, and world. And finally, Shapiro offers Clark's (2008) view that emphasizes the body as more of a 'partner' for the brain and environment in shaping and enacting cognition. Clark's view is often considered exemplary of the extended mind tradition, but again, it too is still called 'embodied' more often than not.

The point of this vast taxonomy of taxonomies is simply to point out that our use of embodiment here will necessarily be controversial, because all uses of the term are controversial. The way we intend to use it (in this section of the text, at least) likely crosses the boundaries of several of these views. When we say that ethical AI (and AI more broadly) needs embodiment, we mean that approaches to AI necessarily require dynamic interactions between perception and action, in a body with certain capacities, which is embedded in a physical environment, but also in a social and cultural context, all of which matter to the kinds of capacities and forms of conceptualization that system will be capable of. In most cases, one can simultaneously hold a view like that of Clark (2003; 2008) or Noë (2009), where the dynamics of the system make it impossible that the brain alone could implement the kinds of representations that most traditional views of cognition demand it to implement, while also holding a view like that of Lakoff and Johnson (1980; 1999) that claims much of the work the body is doing is indeed then encoded in a sort of embodied representation, which enlists and exploits those dynamic systems to further expand our repertoire of abstract concepts, without reducing that notion of embodiment back down to a computationalist-representationalist view.

13. Why We Need Bodies in AI Ethics

We need to attend to bodies in AI ethics because we need to attend to bodies in AI more broadly. There are, as previously discussed, plenty of tangential ways bodies matter in AI ethics, but if we want to do the most interesting kind of AI work, the work of creating a genuine artificial mind, then we need to look to embodiment for all sorts of reasons. No robot could be held accountable for its actions in the interesting way that we hold humans accountable and responsible if that machine is not embodied in important ways. If we accept the premise that the interesting kinds of minds can only emerge through embodied action and perception, then we can start to examine some of the interesting historical arguments (though few and far between) that tie ethics to embodiment.

Let us look again at some of the more traditional claims in ethical AI. Take, for example, the claim (Bringsjord and Taylor, 2012) that some traditional ethical systems are mistakenly left out of the AI ethics discussion, which seems obviously true. In this case, the argument is about the Divine Command Theory (an ethical system in which there are strict laws to follow that are handed down by some god, generally the god of the Abrahamic religions) that, when strictly followed, alleges to give us our system of moral rights and wrongs. Once given the moral system in question, the claim is generally that we can program such a system (be it in explicit action-based rules, or coded into the OS, etc.). If we look across a wide range of arguments in ethical AI, we see this general model emerge time and again, with different large historical ethical systems occupying the space where Divine Command Theory is used here. In many ways, it is the obvious outgrowth of a more general belief that an AI system can be programmed in a box, bedridden, as Daniel Dennett used to call it. (The legacy of Asimov's Laws again!) If GOFAI is a live approach, then a set of easily articulated rules for ethics is no different than those rules being used to govern any other set of behaviour. Mark Johnson, in his book *Moral Imagination* (1993), sums up the landscape well:

> A great many people believe that the way out of our present moral confusions is to get clear about the ultimate moral principles or laws that ought to govern

our lives and to learn how to apply them rationally to the concrete situations we encounter every day. Some people believe that these moral laws come from God, others regard them as derived by universal human reason, and still others see them as based on universal human feelings. However much these people may disagree about the source of moral principles, they all agree that living morally is principally a matter of *moral insight* into the ultimate moral rules, combined with *strength of will* to 'do the right thing' that is required by those rules.

p. ix

This is, precisely, the claim set forth by so many people who claim to have solved the problems of AI ethics and completed the work of hard-coding those ethical laws into any given system. Job done! We can all go home! Except, there are abundant reasons to believe this entire structure of belief about the nature of ethics and the approach to getting them into an artificial mind is deeply flawed and divorced from the reality of our human lives and human morals, which is the only existence proof we have of any such moral system at all upon which to model an artificial system.

There is no settled consensus here, either among ethicists or AI theorists, about the appropriate ethical system, the one that most accurately reflects our world and our understanding of it. There are many live systems of belief here and now, and experts generally disagree about both which system we do use and which system we ought to use. The big three – deontology, utilitarianism, and virtue ethics – generally sit on a tier above the rest, but any introductory philosophy student can rattle off a number of well-established arguments against each one. And while we will return at length to say more about these views (in particular, virtue ethics, which, of the established three, seems both most promising and most daunting with regard to AI, insofar as it may not be a tractable problem that AI is capable of dealing with), here we want to continue focusing on bodies a bit.

In his rebuttal to the GOFAI, rules-based view of ethics he lays out above, Mark Johnson (1993) goes on:

> Something crucial is missing in this widely held conception of morality. What is missing is any recognition of the fundamental role of imagination in our moral reasoning. We human beings are imaginative creatures, from our most mundane, automatic acts of perception all the way up to our most abstract conceptualization and reasoning. Consequently, our moral understanding depends in large measure on various structures of imagination, such as images, image schemas, metaphors, narratives, and so forth. Moral reasoning is thus basically an imaginative activity, because it uses imaginatively

structured concepts and requires imagination to discern what is morally relevant in situations, to understand empathetically how others experience things, and to envision the full range of possibilities open to us in a particular case.

pp. ix–x

The argument being offered here in summary is simply that *no* strictly rules-based system is going to capture the actual ways human beings are moral (and, we would hope, how we can make machines moral in similar ways). In the case of moral imagination, a view which has seen surprisingly little action in AI ethics (and which, we hope, will be taken up seriously in this field in the future), the body plays an almost outsized role in ethics, because of the reliance of this kind of imagination on conceptualization which, in Johnson's case, rests squarely on the shoulders of our particular kinds of bodies in our particular kinds of social and cultural and physical environments.

The argument put forth first by Lakoff and Johnson (1980; 1999) about the body's role in facilitating and enabling the process of conceptualization used by us (and, one would guess, by any biological creature that conceptualizes, as well as any potential non-biological creature that may exist capable of conceptualization, too) is generally referred to as conceptual metaphor theory. It rests strongly on the claim that, 'There is no such fully autonomous faculty of reason separate from and independent of bodily capacities such as perception and movement. The evidence supports, instead, an evolutionary view, in which reason uses and grows out of such bodily capacities' (Lakoff and Johnson, 1999, p. 17). In other words, rather than separate or separable modules of different kinds of processing, with cold reasoning over here and emotion over there, perception one place, action another, and moral reasoning somewhere else altogether, these capacities are all bound up together, and our neural systems, evolved as they are, structure our concepts through our forms of embodiment, reflected in the way our languages work.

We will not attempt to give a complete discussion or critique of conceptual metaphor theory here. Rather, we feel it is sufficient to offer just enough of a background to the view that the claims about embodiment as they relate to AI and ethics can be properly evaluated. It is important to note that Lakoff and Johnson explicitly reject much AI work as a metaphor (the mind as computer metaphor) taken too far, but that shouldn't stop us from applying their claim to work in AI as long as the applications are not contradictory to the initial claims. (In certain research quarters, e.g. AI and cognitive science, the claim about the mind is properly called the computer *hypothesis*. Here, we are focused on the metaphor.) To take a simple example, consider the concepts of *front* and *back*. Lakoff and Johnson point out that these concepts are body-based and emerge from beings (us) whose bodies have fronts and backs (1999, p. 34). They argue that, 'if all beings on this planet were uniform stationary spheres floating in some medium and perceiving equally in all directions, they would have no concepts of *front* or *back*' (34). This is one example of one kind of argument they offer for the ultimate conclusion that all reasoning is embodied, and therefore our reasoning capacities are not accessing some universal, objective truth about 'fronts' and 'backs', but instead things are true by virtue of our shared conceptualization about them (where that conceptualization is indeed shared). In other words: our specific bodies play a large role in how we move through and make sense of the world, as well as how we see and practice rationality and, relatedly, morality.

If we understand the claims made about embodiment broadly as they relate to conceptualization, then Johnson's work applying this to morality is obvious. Why should moral reasoning be exempted from reasoning? It shouldn't, and so it is somewhat surprising that more people in ethical AI have not embraced and explored this view. The last several years have seen an explosion of interest in AI ethics, and there is good reason to believe that the interest will continue to grow. With some luck and careful outreach, the community will seriously examine *moral imagination* as a live option going forward. The benefits of thinking about moral imagination in the context of AI ethics include the fact that the evidence for conceptual metaphor theory is broadly interdisciplinary and fits parsimoniously with both evolutionary theory and what we know about the most current neuroscience. There are downsides of course, including the fact that this version of morality may be intractable with regard to programming and engineering, but these are questions that remain to be explored further.

Part of what seems promising and interesting about moral imagination as a moral framework is that it is bound up with a claim about how we explain our moral decisions via narrative that is at once argumentative (as justification for our actions) and also situated specifically within our social and cultural contexts. It is not a claim about *relativism* (truth is relative to the individual or to a culture), but it is an argument against *pure objectivism* (there is a mind-independent world). It is a moral framework that keeps the human being centred in the analysis, reminding us that what we are, and how we think, are ultimately aspects of our embodied situations in temporal, social, and cultural spaces. Unlike, for example, deontology or utilitarianism, moral imagination allows for all the contradictory rules and beliefs we are known to have at the same time, because it denies that moral deliberation or justification is merely a matter of following rules.

Johnson (1993) offers a telling illustration that describes how moral imagination works. He describes how, as his draft number came up during the Vietnam War, he had to navigate two competing moral obligations: the need to serve one's country, and the deeply held belief that being in Vietnam was a 'grave moral mistake' (p. 185). He lists off all of the relevant arguments for both sides, as well as the possible philosophical considerations of competing moral systems. And then, he recognizes:

> All of this experience. All of this education. All of this knowledge. *And I didn't know what in the world to do!* I had all of the information one could want ... I had all the arguments, philosophical and theological, one could imagine. I had all of the moral education I could handle. I had moral ideals aplenty. I had all of the moral laws I could use, and then some. *And I couldn't decide what was 'right.'*
>
> p. 186, emphasis original

This seems to cut right to the heart of traditional critiques of moral theories, insofar as it recognizes the common way that all theories seem to fail: *there just isn't one simple method for deciding how to act.* And rather than trying to argue for one rule-based theory that seems, on principle, like the ideal case, and programming that into our future AIs or robots, acknowledging the way humans actually seem to reason morally, in all its complicated messiness, is going to go a long way toward making the possibility of moral AI a reality. Johnson summarizes:

> ... a critical moral imagination of this sort ought to be the basis for our moral deliberation and self-understanding. Ideally, moral imagination would provide the means for understanding (of self, others, institutions, cultures), for reflective criticism, and for modest transformation, which together are the basis for moral growth.
>
> p. 187

Insofar as moral imagination is useful for humans, it ought to be useful for AI, too.

Perhaps surprisingly, very little work has been done in philosophy broadly on the concept of moral imagination in ethics, even if we ignore the AI aspect. Outside of Johnson's 1993 book, there exists little work explicitly taking up this notion of moral imagination as it comes to contemporary philosophy via John Dewey. One notable work, though, is Fesmire's *John*

Dewey and Moral Imagination (2003). Drawing from the same well as Johnson's work, Fesmire goes to great lengths to avoid the rigid and, he argues, false framework that leads people to believe they must choose a single moral system to subscribe to in all situations, as if simply being more inflexible with your deontology or utilitarianism will somehow lead to moral clarity and pious moral outcomes.

Fesmire summarizes Dewey's view by claiming:

> Imagination in Dewey's central sense is the capacity to concretely perceive what is before us in light of what could be. Its opposite is experience narrowed by acclimation to standardized meanings.

> p. 65

Those who may rightly cast off GOFAI-style ethical reasoning as impossible in principle but fall back on neural-net style reasoning with a belief that it can produce ethical action may not find much comfort in Fesmire-via-Dewey's understanding of imagination. 'Experience narrowed by acclimation' is a fairly apt description of the ways we tend to train (and then freeze) neural networks. However, combined with Johnson's embodied reasoning, we might see an interesting and unexplored path forward for AI ethics in the future via embodied robotics. Of course, since this concept draws on Dewey, it implies that pragmatism as an approach has something valuable over the alternatives, but it doesn't dismiss the value of those alternatives entirely.

Unlike the situation we find ourselves in when we choose to align our morals with Kant or Mill, Dewey (and by extension Fesmire here, and Johnson above) accepts that the rules and principles of competing moral stories may all have something worth retaining. As Fesmire puts it:

> On a pragmatic view, principles and rules *supplement* philosophical ethics; they do not constitute it. They ideally play an orienting and economizing role in everyday decisions, so they must be conserved in some form. For ethics to be revitalized, what needs to be challenged is the belief that principles, rules, and the systems they comprise must constitute the tethering center of either ethical theory or practice. This dogma can obstruct experience rather than ameliorating it, yet it is tenaciously held to define ethics itself.

> p. 3

This kind of claim is quickly dismissed in most AI work, because it acknowledges (and perhaps even embraces) a messiness, a flexibility, and a lack of rigidity with regard to rules that makes for unpredictable outcomes. But of course, our own moral deliberations are unpredictable, and some philosophers have gone so far as to place the weight of the value of the deliberation itself on the fact that deliberation must be present. As opposed to a simple calculation with a clear and decisive outcome, moral deliberation is messy. Necessarily so, some would argue.

Again, this notion of moral imagination is a promising route toward thinking in different possible futures about ethics and AI, but it would be nonsensical to divorce it from its practical and historical roots in embodiment. Dewey, like many wise philosophers before and after him, laments how our historical split between body and mind has affected our ability to do philosophy across a wide number of domains. He notes, 'the division in question is so deep-seated that it has affected even our language. We have no word by which to name mind-body in a unified wholeness of operation' (Dewey, 1928). This presages the attempts of later phenomenologists, like Merleau-Ponty and Heidegger, to create new words to capture that notion of mind/body as a singular entity. (These attempts have never stuck, likely for a number of reasons, most of which trace back to the very metaphysical implications of mind versus body that are at stake in this historical moment, where we are still debating the meanings of these terms even while we tinker in our neural networks and robotics labs in an attempt to recreate the thing we haven't yet understood.)

Dewey was aware of the implications of this mind/body split for ethics, and it's unfortunate that little work in embodied ethics has been done in light of our AI projects. If AI isn't at least partially an attempt to understand our own natures, it's hard to know what other hands-on project could be. Dewey rightly argues:

> I do not know of anything so disastrously affected by the tradition of separation and isolation as is this particular theme of body-mind. In its discussion are reflected the splitting off from each other of religion, morals and science; the divorce of philosophy from science and both from the arts of conduct. The evils which we suffer in education, in religion – for example the fundamentalist attack about the evolution of men rests upon the idea of complete separation of mind and body – in the materialism of business and the aloofness of 'intellectuals' from life, the whole separation of knowledge and practice – all testify to the necessity of seeing mind-body as an integral whole.
>
> p. 27

Of course, Dewey wasn't concerned with AI in 1928, but his overall worry about the practical implications of conceptually splitting mind and body were both accurate and prescient for much of the last century of culture wars between the arts and the sciences, at least in the most public spheres. But what matters for us here and now is that we have an abundance of philosophical (and practical) arguments showing us that considering mind and body one entity has implications well beyond any single way of making sense of the world. Here, the possibility of engaging imaginatively with our moral dilemmas rests, in serious ontological ways, on our making sense of how bodies and environments (social, political, cultural, physical) create the concepts we use to engage in those moral deliberations, always entangled and never in some free-floating, Cartesian mind-soup, easily plugged into a bedridden computer system that might spit out convenient claims that we might then label 'ethical' or 'not'.

14. The Social Life of Embodied Ethical AIs

Here we have been discussing the relationship between embodiment and ethics, both broadly and, one hopes, with the goal of applying this work to the future of artificial intelligence. (We've tried everything else; why not give this a shot?) We have spoken of the importance of the body in its relation to the environment for generating abstract concepts (both moral and not) but have only really hinted at the ways the social world will matter. And the tie between embodiment and the social world is perhaps still less clear. Here, we will discuss some ways that embodiment, social engagement, and ethics may be tied together such that any future attempts at AI must go beyond merely being actually embodied in the physical world, they must also include embedding that embodied system in our shared human social world. (And while there remain many untapped resources in historical philosophy for future AI, it remains to be seen if these conclusions can be enacted in any fruitful way, or if instead they indicate the impossibility of the task.)

Think back to some of our earliest debates in AI: the feasibility of the Turing Test as a valid measure of artificial intelligence; the proper approach to building the structure of intelligence itself (Newell and Simon's Physical Symbol System Hypothesis versus some neural network versus embodiment); the Symbol Grounding Problem. All of these debates, and many others, focused around the nature and structure of language in one way or another. While philosophers wondered if the Turing Test could ever possibly produce meaning and understanding because it is just language in the void (Searle, 1980), the long history of the function of language and speech were sometimes forgotten in these debates. Philosophers engaged with Turing and Searle but often excluded Wittgenstein and Dewey. For Dewey, for example, like Merleau-Ponty years later, sounds and vocalizations alone could never be speech or language, since speech and language were inherently social technologies (Dewey, 1928; Merleau-Ponty, 1945). The idea that language and thought were the same, and then you opened your mouth (or typed your text) to communicate those ideas, was a gross misunderstanding of the social and embodied aspects of language. Dewey claimed that 'inclusion of objective social consequences is what transforms sounds into speech or language' (p. 37). Of course, Dewey traces this back to our grave error in splitting body and mind from the environment, because it means we ignore not just the behavioural aspects of language, but the social as well.

Of course, pragmatism as an approach has been offered very little attention historically in the debates about artificial intelligence, and some may argue that's been to the detriment of the AI program. But when it comes to thinking about ethics, there is so much richness here that has barely been explored, that there is reason for some optimism with regard to philosophy's role in the future of thinking about AI and ethics.

As mentioned earlier, there have been explorations into the potential relationship between things like empathy and consciousness (Thompson, 2001; 2007). And there have been some attempts to tie this back to ethics, particularly with regard to AI (Torrance, 2008; 2014). Perhaps the reason that many don't want to accept empathy as a prerequisite for consciousness, and embodiment as a prerequisite for empathy, is that it seems to wed one to the underlying biology and therefore discount the possibility of ethical embodied AI at all. And pragmatism and phenomenology would seem, on the face of it, to discourage us in exactly this direction. But the lessons from these approaches may still hold value for us, if we care to find it. They offer us paths through the endless attempts at finding clear rules for conduct, instead demanding action in the world and with other beings, the ultimate goal of AI. Fesmire reminds us:

> On Dewey's view, direct valuing such as empathy is complemented and expanded in deliberate, practical reflection, which requires tapping possibilities for action – especially through communication and in dialogue – and forecasting the consequences of acting on them. Both aspects of imagination, empathy and creative tapping of possibilities, operate simultaneously. This focuses deliberation concretely on the present yet expands attention beyond what is immediately experienced so that the lessons of the past, embodied in

habits, and as-yet-unrealized potentialities 'come home to us and have power to stir us'.

p. 67

Empathy, perhaps possibly only in certain kinds of embodied social creatures, ties itself up with action and reflection in a way that enables ethical behaviour. But communication and dialogue, inherently social actions, play a big role. So our GOFAI systems, by this understanding, were never going to accomplish anything. Our neural networks wouldn't fare much better; but a fully embodied, social being (in whatever material form that needs to take), might.

Again, embodiment is a fraught concept, and there may be more than one way to make an artificial embodied being capable of the kinds of conceptualization necessary to use the moral imagination needed to engage with the world ethically. There has been plenty of work done in neuroscience in recent decades that offers some hints at how the social world impacts our cognition in concrete ways, but it overwhelmingly favours the understanding of our cognitive systems as full-bodied, embedded, social systems. For example, much of the work done by Antonio Damasio in the last twenty-five years demonstrates his primary argument that social reasoning is particularly bound up with emotion in a necessarily messy neural system that denies the traditional split between rationality and emotion, demonstrating instead the ways that emotion is a necessary precursor to clear and productive reasoning (Damasio, 1994; 2003). Damasio highlights particular neural regions (in this case, the ventromedial prefrontal cortex, or VMPFC) that he says are implicated in social reasoning that requires emotional investment. He offers case studies from his own medical practice, placing his patients in a historical context that aims to show how the particular brain area in question, when damaged in certain ways, hinders time-pressured social reasoning. His patients show a range of deficits, from an inability to hold down a job or marriage to just being incapable of choosing an appointment time, given two seemingly equally appealing offers.

Some of Damasio's work may be surprising to many working in AI, because what his patients appear to demonstrate is a kind of pure rationality – detached and objective reasoning the likes of which would make Kant blush. But rather than being the most successful rational actors that economists and other social scientists sometimes seem to think we ought to aspire to be, most of Damasio's patients instead cannot function, likely because our values and desires are deeply bound up with our emotional attachments, often in non-obvious ways. Damasio recounts the story of a patient who was so emotionally detached that he became like Buridan's Ass, unable to choose a time for a follow-up appointment and effectively frozen in endless deliberation as a result. He says:

> I was discussing with the same patient when his next visit to the laboratory should take place. I suggested two alternative dates, both in the coming month and just a few days apart from each other. The patient pulled out his appointment book and began consulting the calendar. The behavior that ensued, which was witnessed by several investigators, was remarkable. For the better part of a half-hour, the patient enumerated reasons for and against each of the two dates: previous engagements, proximity to other engagements, possible meteorological conditions, virtually anything that one could reasonably think about concerning a simple date . . . he was now walking us through a tiresome cost-benefit analysis, an endless outlining and fruitless comparison of options and possible consequences . . . we finally did tell him, quietly, that he should come on the second of the alternative dates. His response was equally calm and prompt. He simply said: 'That's fine.'
>
> 1994, p. 194

Of course, this is a case where nothing is really at stake, but many of Damasio's patients with VMPFC damage lost their livelihoods, their families, and their freedoms. Damasio points out that this sort of behaviour really highlights the limits of pure reason, an idea he thinks philosophers and psychologists alike have been deeply mistaken to value or desire.

Importantly, much of the neuroscience work Damasio offers involves uncovering complicated neural and bodily systems that seem to have evolutionary roots (accounting for instinct, for example), but have evolved the ability to be tuned through and to social influence, over the development of the individual, accounting for things as different as that gut-wrenching feeling we get when we've committed a social faux pas, like forgetting the name of a friend, and the deep, bodily unease we feel when we are introduced to some grave moral subversion. In this way, Damasio manages to tie together not just felt bodily experiences that are likely to be necessary for reasoning in general, but also the ways our moral reasoning relies on those systems and experiences. And in this case, the neuroscience accords well with what many philosophical theorists have argued: embodiment matters, and it matters that the body is embedded in a social context, and develops in that social context, in order to achieve the kinds of reasoning (both moral and rational) that we claim we want to achieve with strong AI.

Of course, plenty of ethical systems recognize the importance of the social world. All social contract theories are relevant here and may have something to offer us as we consider moving ahead in building ethical AI. And while Rawls's veil of ignorance may play out slightly differently than Scanlon's version, this kind of contractualism is only promising for us if we can already imagine an AI with experiences of the sort that we consider to be worthy of moral consideration in the same way other members entering genuinely into the social contract are.

15. Artificial Phronēsis

On one hand, consciousness is the easiest thing to experience, since consciousness is necessary for rich fully experienced perceptions, of the kind that we humans have regularly. Thus, the vast majority of those who read this paper are likely to be conscious (at least some of the time) and are fully aware of what it is like to be conscious. On the other hand, consciousness is a phenomenon that is notoriously difficult to fully explain in a way that would allow a computer engineer to build a conscious machine, even after there have been many decades of heroic attempts to do so in our recent history.

Much work can be done in AI and robotic ethics that does not require that the systems in question have any consciousness at all (or at least none that we humans have to acknowledge and accept). There are three levels of artificial agents that will be discussed here. They are classified by their ethical abilities, or lack thereof: ethical impact agents (EIA), artificial ethical agents (AEA), and artificial moral agents (AMA).

Ethical impact agents (EIA) need no explicit claims to consciousness to do what they do. These systems are notable only in that in their operations, they exhibit some autonomy and have the ability to impact human agents in ways that have ethical concern. For instance, an autonomous car that, during the course of its autonomous driving operation, impacts and kills a pedestrian in an accident (such as the accident that happened on 18 March 2018 in Tempe, Arizona involving a pedestrian and a self-driving Uber car) does so completely unconsciously. The System's actions are produced by its sensors and actuators working in accordance to a program, but it has no conscious experience of the situation at hand, suffers no moral remorse after the accident, and has no emotional reaction while the event is unfolding. While we might ask plenty of questions about the safety of the autonomous car, no one blames the car itself in a conscious moral sense. That moral condemnation and all questions of legal responsibility are reserved for the emergency backup driver present in the car, and the company that built the car and is testing it on public roads. Thus, there is no particular need to refer to consciousness when dealing with questions on how to ethically design and deploy EIAs.

Artificial ethical agents (AEA) are the next step up and differ from EIAs only in that they have explicit ethical considerations programed into their operation. Building on our earlier example, an autonomous car that was programmed to take into account some ethical calculus of value when deciding whether to risk occupants of another vehicle or its own occupants to increased risk during an otherwise unavoidable crash would be an AEA. At first look one might think that an AEA muddies the waters a bit and that the machine itself might deserve moral blame or legal responsibility, but that is just a trick of the light. The moral blame and responsibility for any adverse consequences is still fully borne by the human agents that built, deployed, licensed, and operated the vehicle. The key is that the AEA never itself choses its own ethical standards, instead they are chosen and programed in by human agents, who therefore assume any blame or responsibility for any ethical decisions made by the system that they design and/or deploy. The machine is not conscious of any of the events that occur based on its operations, even the ones that look to an outside party as if they were a conscious ethical choice.

It is only when we get to the level of the *artificial moral agent* (AMA) that consciousness may play an important role. An AMA would have the ability to choose ethical behaviours that are appropriate to the situation at hand in a way that exhibits a form of practical reasoning similar to what can be seen in competent ethical reasoning found in most human agents. This means that the system either is a conscious moral agent or is functionally equivalent to one.

Another way to say this is to claim that the system displays *artificial phronēsis* (AP). Of course, this concept needs a lot more explanation and that is what the rest of this chapter will discuss. However, at this point in the discussion we can make the conjecture that, while consciousness is not required for EIAs nor for many types of artificial ethical reasoning agents, it may play an important role in the development of much more sophisticated AMAs that would be much more useful in interacting with humans in complex social situations that tend to be bounded by shifting ethical norms.

'Phronēsis' (pr. fro *Nee* sis) is a term that many are not familiar with outside of philosophy and the word can seem a little off-putting. However, if one's goal is to create AI and robotic agents that have the capacity to reason intelligently about ethical situations, then understanding this technical term will reward those who try, given that it is so relevant to the understanding of ethics.

'Phronēsis' has an ancient pedigree and has come down to us largely through the tradition of virtue ethics as the skill of being able to *live well*. 'Phronēsis' refers to the practical wisdom that a conscious moral agent uses when she is confronted with a difficult moral or ethical problem and attempts to overcome these difficulties in an intelligent, moral manner. Given that ethical problems are often so complex, no set of preconfigured answers will suffice to solve the problem (as discussed in section 14, above). This means that learning and creativity are the hallmarks of a phronētic agent.

AEAs might be successfully designed taking one or more ethical schools of thought into account. One might design an autonomous system that makes ethical decisions based largely on applied utilitarian or Kantian based calculations or rules, models of human moral psychology, human religious traditions, or even on the three laws of robotics developed by Isaac Asimov. While one could make AEAs using any of these methods that might be useful in certain circumstances, they will all fall far short of an AMA with artificial phronēsis.

It is understandable that systems designers will either want to ignore ethical reasoning entirely, attempting to avoid the construction of even EIAs. The slightly more adventurous will attempt to apply the more computationally tractable rule-based ethical systems that could result in useful AEAs. Why get involved with AMAs that require something like artificial phronēsis to work correctly? To succeed at that may require solving problems in artificial consciousness, artificial emotion, machine embodiment, etc., all of which may be computationally intractable. Let's look at what might be the reward for pursing the more difficult problem of building AMAs with artificial phronēsis.

Robot Rights

If we were able to create a machine that displayed artificial phronēsis, then would that machine have a claim to rights as well as responsibilities? This is a question that has been addressed by the philosopher David Gunkel (2019). His argument proceeds as follows: S1 'Robots can have rights' or 'Robots are moral subjects.' S2 'Robots should have rights' or 'Robots ought to be moral subjects.' The question 'Can and should robots have rights?' consists of two separate queries: 'Can robots have rights?' which is a question that asks about the capability of a particular entity. And 'Should robots have rights?' which is a question that inquiries about obligations in the face of this entity.

Potential outcomes of this are Not S1 and Not S2: 'Robots cannot have rights. Therefore, robots should not have rights.' S1 and S2 'Robots can have rights. Therefore, robots should have rights.' S1 and not S2 'Even though robots can have rights, they should not have rights.' Not S1 but S2 'Even though robots cannot have rights, they should have rights.' Although each modality has its advantages, none of the four provide what would be considered a definitive case either for or against robot rights. Gunkel (2007), thinking otherwise, does not argue either for or against the is-ought inference but takes aim at and deconstructs this conceptual opposition. He does so by deliberately flipping the Humean script, considering not 'how ought may be derived from is' but rather 'how is *is* only able to be derived from ought'. At this point in his analysis, Gunkel refers to Emmanuel Levinas who, in direct opposition to the usual way of thinking, asserts that ethics precedes ontology. In other words, it is the axiological aspect, the 'ought' dimension, that comes first, in terms of both temporal sequence and status, and then the ontological aspects follow from this decision.

The Role of Artificial Phronēsis in Automated Ethical Reasoning

Artificial Phronēsis (AP) is the claim that phronēsis, or practical wisdom, plays a primary role in high level moral reasoning and further asks the question of whether or not a functional equivalent to phronēsis is something that can be programed into machines.

If we want AI systems to have the capacity to reason on ethical problems in a way that is functionally equivalent to competent humans, then we will need to create machines that display phronēsis or practical wisdom in their interactions with human agents. This means that AP is one of the highest goals that AI ethics might achieve. Furthermore, this will not be a trivial problem since not all human agents are skilled at reasoning phronētically, so we are asking a lot from our machines if we try to program this skill into them. On top of this, the most difficult problem is that achieving AP may first require that the problem of artificial consciousness is solved, given that phronēsis seems to require conscious deliberation and action to be done correctly. Even so, the achievement of AP is necessary, since moral and ethical competence is required in order to develop ethical trust between the users of AI or robotics systems and the systems themselves. Achieving this will make for a future that humans can be comfortable inhabiting and not feel oppressed by the decisions made by autonomous systems that may impact their lives.

Artificial Phronēsis: a Manifesto

AP is a new concept, but it is gaining some attention. What follows are some statements to help define this new, interdisciplinary area of research.

AP claims that phronēsis, or practical wisdom, plays a primary role in high level moral reasoning and further asks the question of whether or not a functional equivalent to phronēsis is something that can be programed into machines.

AP is a necessary capacity for creating AMAs, however, the theory is agnostic on the eventuality of machines ever achieving this ability, but it does claim that achieving AP is necessary for machines to be human equivalent moral agents.

AP is influenced by works in the classical ethics tradition, but it is not limited to only these sources. AP is not an attempt to fully describe phronēsis as described in classical ethics. AP is not attempting to derive a full account of phronēsis in humans either at the theoretical or neurological level. However, any advances in this area would be welcome help.

AP is not a claim that machines can become perfect moral agents. Moral perfection is not possible for any moral agent in the first place. Instead AP is an attempt to describe an intentionally designed computational system that interacts ethically with other human and artificial agents even in novel situations that require creative solutions. AP is to be achieved across multiple modalities and most likely in an evolutionary machine learning fashion. AP acknowledges that machines may only be able to simulate ethical judgement for quite some time and that the danger of creating a seemingly ethical simulacrum is ever present.

This means that AP sets a very high bar to judge machine ethical reasoning and behaviour against. It is an ultimate goal, but real systems will fall far short of this objective for the foreseeable future.

Dewey on the role of Phronēsis in Conscious Thought

If phronēsis was simply a concept only from ancient philosophy, it would be of limited value to the project of AI. But it has evolved over time and one very interesting development of the concept came from the philosopher John Dewey (1928/1984; 1998a; b). Unlike the ancient philosophers who seem to use phronēsis to denote a capacity that only the most highly intelligent philosophers possess, Dewey greatly expands the concept to one that plays a central role in all manner of reasoning due to the fact that once you try to apply any science or skill you necessarily enter into the social sphere and successfully operating there requires phronēsis. If Dewey is correct, then phronēsis is part of what makes many of us competent, conscious, and conscientious beings. It follows then that AP is either essential for the creation of conscious machines, or vice versa.

Ethical Trust of AI and Robotic Agents

Trust and AI and robotic agents is its own complex topic, but here let's limit our discussion to phronētic trust. Our AP agents will need to convince us that they have our moral character in mind when they are dealing with us and will make decisions accordingly. We learn from each other how best to develop our own moral character, so these machines will need to participate in that important social process as well. In some sense, they must be able to serve as a phronēmon, or moral teacher. In these kinds of relationships with our machines, we have to be warranted in reasoning that the machine we are trusting has a good 'character' that is deserving of our trust. Without a sufficiently developed AP, then, this will be impossible and there will be no good reason to try to build artificial ethical agents and we will need to limit the applications in which we employ AI to only those with no ethical impact.

16. How Can We Ethically Trust AI and Robotic Systems?

Trust is a rich field of philosophical study that spans multiple definitions of trust, deciding who to trust, and even what kinds of technologies to trust (See, Simon, 2020). When it comes to AI and robotics there are a few emerging theories on when we are justified in trusting AI and robotic systems. Mark Coeckelbergh (2012) argues that there are at least two distinct ways we can look at trust in AI and robot systems. The first is in a 'contractarian-individualistic' way where we might enter into trusting relationships with these systems in a similar way in which we enter into trusting relationships with corporations or other legal entities when we rely on their products or services. These relationships may be more or less prudential, but they are a long-established practice in our societies. Taddeo and Floridi (2011) have developed some of the nuances that will need to be considered as these systems become more competent and the trust we place in them becomes more meaningful.

Coeckelbergh's second category is the phenomenological-social approach. This is the kind of deep trusting relationships that humans commonly enter into with each other that develops into social bonds and creates deep and meaningful relationships. Coeckelbergh is sceptical that this can be achieved in AI and robotic systems in any way that is not 'virtual-trust' or 'quasi-trust' where these systems might be good at playing us in social games and garner our trust, but we do so at our own risk. While we may be fooled into thinking we are in a relationship of ethical trust with the AI or robotic system, in fact we are not, and in some situations this could be dangerous to the undiscerning human agents.

Grodzinsky, Miller, and Wolf (2011, 2020) provide a system that might mitigate this problem through the development of a new concept called 'TRUST' thought of in an object-oriented way in which TRUST, 'will include traditional, face-to-face trust between humans, and "TRUST" will also include electronically mediated relationships and relationships that include artificial agents'. Here we clearly define a new system of trust in machines that fits under a new heading category that also contains our already well-developed notions of human to human trust and human to corporation trust, etc. Only if we wanted machines to join us in the phenomenological-social trust would we really need AP. At that level we would join them in a new kind of society. That is a very interesting eventuality to contemplate but it is also one that would come with rights and responsibilities that would be bidirectional, so we should proceed with caution, and if it turns out that machine consciousness is technically infeasible, then it is something that we could not approach at all.

Sullins (2020) argues that we must consider what he calls "ethical trust" where we ask if trusting a robotic system is in itself ethical, whether or not the trust is actual trust or some new form of trust, *eTrust*, for example. An example would be that one user's trust that a cleverly designed humanoid robot could manipulate human psychology and manipulate users like a master salesperson might, and this could impact others who are not in direct interaction with the robotic system.

17. Conclusion

This chapter has tried to imagine some of the promising ways philosophy, coupled with cognitive science, can offer us guiding pathways to building ethical artificial intelligence. There is no commitment here that claims there is one right way to do this, or even that it's necessarily possible, but there is so much richness yet to explore that it seems premature to imagine we've already solved this problem. There has been a big emphasis on embodiment, and how that plays out across a number of possible ethical frameworks. Very few people working on the interesting, mad-scientist kind of strong AI are starting from the embodied framework at all. Embodied AI is hard. It's expensive. It demands the kind of social and physical engagement with a real world that makes it very hard to theorize without building. Again, hard and expensive are disincentives to taking any of the approaches laid out here. Indeed, even if we divorce the ethical question entirely, if we take embodiment to be a necessary condition for *mindedness*, we're still stuck with the difficulty and expense. Still, even though we're trying to recreate in our own lifetimes a project that took nature, with all her resources, billions of years to achieve, it's still worth a shot.

Conclusion: Whither the AI Wars?

The AI Wars of the latter twentieth century have gone quiet now. Philosophers, for the most part, no longer object to AI, except from the various ethical points of view explored in Chapters 8 and 9. Few talk about the logical possibility of AI, and debates about architecture are generally left to practitioners. The war about aboutness has morphed into the war about consciousness. Many philosophers are deeply engaged in this debate, but it is now motivated more by neuroscience than by AI. The frame problem remains unresolved, but lurks everywhere, unrecognized.

Are these twenty-first century conflicts about consciousness and ethics further 'AI wars'? Only time will tell.

AI researchers carry on developing systems, now in the twenty-first century, without philosophers nipping at their heels. But all the same, no one won the twentieth-century AI wars. So why did the wars go quiet? Partly because AI did not deliver robots with human-level general intelligence, as it promised. Perhaps the many philosophers who objected to the very idea of AI can feel vindicated: AI researchers have not succeeded in solving the problems of machine mental semantics or the aboutness of computer symbols, they never seriously addressed the problem of machine consciousness, they never succeeded in getting machines to grasp what is relevant to what. So, AI failed. Very tidy. However, it is now clear, at least to some of us, that if AI is going to deliver our intellectual equals or betters – in large and natural domains (i.e. real life) as opposed to tiny, circumscribe domains – these problems will have to be addressed.

Advances in AI since 2000 suggest, interestingly, that the intelligence might come first, followed by insights into mental semantics, relevance, etc.

Circumscribed domains may grow incrementally until they start to overlap and merge. This path might be the only path.

A deeper concern, however, here is that these 'AI problems' remain today *human problems*. We have no idea how human or animal minds have mental content, how human or other animals recognize and keep track of what is relevant to what (when they do), and of course, consciousness remains an impenetrable mystery. So, it is not as if the AI researchers can just import what the psychologists and neuroscientists have figured out and call it a day. Psychology, despite the progress of neuroscience, has accomplished no more than AI has.

But has AI really failed? It is not an old research topic (compare it to physics), and what it is researching is the most complicated collection of phenomena in the known universe: the human mind. So, what may appear to be failure may in fact be progress. We have to take seriously that AI has proceeded to develop and grow to the point where, today, nearly all humans feel its impact. Pro-AI researchers of all stripes can perhaps glimpse a light at the end of the tunnel.

The same cannot, however, be said of philosophy. Having raised, as it always does, deep and disturbing problems, cutting to the heart of the matter, it has again failed to provide any answers. Philosophy never has supplied any answers to the deep questions it raises, not once in its millennia long existence – at least not answers that enjoy any general agreement. There is no general agreement on semantics or relevance, for example. There is no general agreement about consciousness or ethics. So, it is here that philosophy takes its final shot: a philosophical machine as intelligent as a human is also not going to provide any answers. This could remain true regardless of how smart or ethical the machines become. One can imagine a future ruled by profoundly intelligent machines, who, like us, are stuck wondering about the questions first raised in the twentieth century's AI Wars.

Notes

Chapter 1

1. Lucas published his paper in 1961. It uses an important theorem from logic to argue that computers will *never* have the intelligence of human mathematicians. Lucas's attack is examined in the First War.
2. See https://www.epa.gov/greenvehicles/greenhouse-gas-emissions-typical-passenger-vehicle. Retrieved June 2019.

Chapter 2

1. At this point, the reader no doubt expects a lengthy discussion on criticisms and counter-criticisms about the Turing Test. But we are skipping these because to discuss them in any detail would be seeing the Turing Test as something that we argue it is not.
2. Do-Much-More was the best performing computer entry in 2009; see https://aidreams.co.uk/forum/index.php?page=Do-Much-More_Chatbot_Wins_2009_Loebner_Prize#.Xl2kXullDCI.
3. For a transcript, see https://www.scottaaronson.com/blog/?p04&paged=24.
4. The Third and Fourth Wars, especially, can be easily described as instances of philosophers making just this kind of mistake. We stress, however, that from another perspective, the problems these philosophers raised remain to this day, for they are profound conceptual problems that apply even to humans.

Chapter 3

1. For a tiny sampling of such progress, see Bellmund et al., 2017, Buzsáki and Llinás, 2017, Carlén et al., 2017, Sousa et al., 2017. However, it remains quite obscure how neurons coordinating their waves of firings in human brains constitute or implement thinking of, say, our dogs.
2. There are still attempts by philosophers to get rid of the most difficult aspects of our inner mental lives. Two well-known examples are Dennett's 1991 book *Consciousness Explained*, and Metzinger's 2003 *Being No One*. Dennett's book tries to rid the world of consciousness, while Metzinger's book attempts to rid us of our selves. There is a large literature on both books, attacking and

defending. To be fair, Dennett means a rather particular thing by 'consciousness' and Metzinger means a rather particular thing by the 'self'; see Chapter 7 for a discussion of these issues. As with similar attempts by the Logical Empiricists, consciousness and the self survived both Dennett and Metzinger due to the fact that the reality of consciousness and selves is impossible to deny. Dennett and Metzinger both accept this impossibility, but seek to explain it with the claims that consciousness and the self are inevitable and unavoidable *illusions*. Perhaps. But perhaps some things, even very important things, simply cannot be explained.

3. Yes, computers have to be implemented in something physical for us to use them. Macs and Dells are physical machines, after all. But as stressed, the details of the physical implementation are not important; this is why they can vary considerably: think of an electronic calculator versus an abacus, for example. All that matters is that there be *some* physical manifestation of the virtual machine.

Chapter 4

1. The 1956 Dartmouth Summer Research Project on Artificial Intelligence (its official name) is usually regarded as the seminal event of artificial intelligence. It was held at Dartmouth College, Hanover, New Hampshire. More can be found at various places on the web, including Wikipedia.

2. Many would say that there are more than four contenders. For example some include artificial life as one contender (we do not, partly because progress in ALife is so slow), some separate embodied cognition from situated cognition (we think they are usefully combined), some include unusual neurocomputational architectures such as Edelman's Neural Darwinism (we exclude these because they are mostly speculative at this time; see Edelman, 1987).

3. See Fodor, 1987, 2000; Searle, 1980, 1990; Patricia Churchland, 1996; Paul Churchland, 1989, 1995; Clark, 2008, 2013; van Gelder, 1995.

4. Formally, a Boolean function is a function $f: \{F, T\}n \rightarrow \{F, T\}$. Read this as "$f$ is a function from a string of F's and T's n characters long to either F(alse) or T(rue). Usually, 0 represents F and 1 represents T, so this definition is most often written as $f: \{0, 1\}n \rightarrow \{0, 1\}$.

5. The qualifier 'in principle' is important. A Turing machine takes as input arbitrarily large strings, processes them, and returns an output. But this is impossible for any real, physical machine, even one the size of our universe, because our universe is finitely big (so the cosmologists tell us these days). That is, a Turing machine can take as input a string far larger than the number of elementary particles in the universe. Of course, such a machine exists only in the abstract, i.e. only mathematically.

6. The original linguistic turn occurred in the early twentieth century as part of the Logical Empiricism of the Vienna Circle. The idea was that all philosophy problems could be solved if we could just get clear about how we were using language; and then repair the incorrect uses. If we used language clearly enough, if we used an ordinary language (like German) constrained by an ideal language (like classical logic), then the philosophy problems that have plagued us since before the Pre-Socratics would melt away. (The terms 'ordinary language' and 'ideal language' are due to the Empiricists.) The AI linguistic turn actually followed the original in many ways. One bit of evidence for this is that we say we program computers using computer *languages* - like Python, Java, C++, Lisp, FORTRAN, COBOL, etc. But by no remotely reasonable criteria is a computer language like a natural language like German. Natural languages are far, far stronger and more flexible. And it is worth noting here that no logic, classical or otherwise, is really a language. The expressive power of any natural language dwarfs that of any logic. A consequence of this is that it is impossible to capture a natural language directly in a logic. Throughout the history of AI and the recent history of logic, many efforts, both in philosophy and in AI, to come up with a logic that did capture the expressive power of a natural language were attempted. They all failed.

7. That this language is innate in human beings was a thesis argued for by Chomsky (1957).

8. This is a more understandable, less cryptic, version of Newell and Simon's Physical Symbol System Hypothesis (Newell and Simon, 1976). For more on this hypothesis, see the Third War: Mental Semantics and Mental Symbols.

9. The old enthusiasm for language-like symbols has been replaced by many new enthusiasms for such things as case-based reasoning, neural nets, semantic ontologies, object-oriented languages, and a host of new logics and logical techniques.

10. The hand-coding problem was, of course, not the only problem raised by critics of the symbolic processing movement. Much criticism also focused on the problem of *aboutness*: what were the symbols about in the world? This is addressed in Chapter 5. Another question was how the computer exchanged information with the world. This was addressed under the rubric of 'Embodied, Situated Cognition'. See below. Many philosophical attacks on AI, e.g. H. Dreyfus' popular *What Computers Can't Do* (1972), combined elements of all of these criticisms.

11. Some philosophers thought the AI hypothesis was obviously false; some thought it was also *offensive*. Here's a quote, which we will just let speak for itself: '[AI]'s real significance or worth [lies] solely in what it may contribute to the advancement of technology, to our ability to manipulate

reality (including human reality), [but] it is not for all that an innocuous intellectual endeavor and is not without posing a serious danger to a properly human mode of existence. Because the human being is a self-interpreting or self-defining being and because, in addition, human understanding has a natural tendency to misunderstand itself (by interpreting itself to itself in terms of the objectified by-products of its own idealizing imagination; e.g. in terms of computers or logic machines) – because of this there is a strong possibility that, fascinated with their own technological prowess, moderns may very well attempt to understand themselves on the model of a computational machine and, *in so doing*, actually make themselves over into a kind of machine and fabricate for themselves a machinelike society structured solely in accordance with the dictates of calculative, instrumental rationality' (Madison, 1991).

12. It isn't clear why, other than lingering Behaviourism, both the embodied cognition and dynamic system researchers are so keen on getting rid of representations. Representations are the key to figuring out cognition and thinking – after all, when we think, we think about things. But keen they are, as this quote demonstrates: '. . . Dietrich and Markman have argued in string of a papers . . . that cognitive scientists should be unified in their acceptance of the necessity of representational explanations of cognition . . . Their main point is that antirepresentationalism is a nonstarter. (The point of this book is that they are wrong about that.)' Chemero, 2009, pp. 52–3. Getting rid of representations won't free cognitive science from some mythical chains that are holding it back; instead it would negate the only progress we've made. One cannot be blamed for thinking that both embodied cognition types and dynamic systems types do not *want* human thinking to be explained, nor do they want to share the planet with intelligent computers.

13. Lubben, Alex (19 March 2018). 'Self-driving Uber killed a pedestrian as human safety driver watched.' *Vice News. Vice Media.* Actually, 'self-driving cars' have killed several people, so far: https://en.wikipedia.org/wiki/List_of_self-driving_car_fatalities. Retrieved June 2019.

Chapter 5

1. One conclusion to draw here is that there are only changing appearances; there is no ultimate unchanging reality behind these appearances. The changing appearances merely supervene or stack up, one kind on top of another kind (e.g. cells on top of organelles, on top of macromolecules, on top of smaller molecules, on top of atoms, and ever downward). The 'reality behind appearances' then becomes *relative*: the organelles are the unchanging reality behind changing cells, and so on, down the ladder. A

related difficult issue is that to say something is changing, logically requires *no change*, for something has to survive the change – otherwise, there would be no change at all, but rather something entirely new would just take the place of something old.

2. This claim about communicating with Aristotle entails that Thomas Kuhn was partially wrong in this famous book *The Structure of Scientific Revolutions* (1962). Kuhn argued that science progresses, in part, by huge paradigm shifts, and that those in one paradigm cannot successfully communicate with those in another paradigm. Darwin's theory of evolution was just such a shift. So, biologists and naturalists before Darwin could not successfully communicate with biologists or naturalists after Darwin. The former held that Earth was young, that life was not changeable, and that it had been created. The latter held (among other things) that Earth was old, life changeable, and it was not created. Of course, one can agree with Kuhn about the importance of paradigm shifts in science without agreeing with him that such shifts block shared meaning and conversation.

3. Edited from *Time Enough for Love* by Robert Heinlein.

4. For Hawking, see http://www.bbc.com/news/technology-30290540, and for Gates, see http://www.bbc.com/news/31047780. Gates does think the threat is not imminent.

5. For a dissenting view, see https://www.technologyreview.com/s/425733/paul-allen-the-singularity-isnt-near.

6. The intension of a term is the collection of attributes belonging to all and only members of the set of objects satisfying or having those attributes. To see this better, contrast 'intension' with 'extension'. The extension of the term 'red' is the set of red things in the world. The intension of the term 'red' is the attribute or property something must have to be in the set of red things – specifically, being red.

7. Some might think that calculators, mousetraps, and thermostats are not computers, and so are not relevant to the argument that computers are meaningless. But calculators, mousetraps, and thermostats are computers. Here, the historical discussion gets very technical, involving such notions as Turing machines, digital versus analog processing, continuity, and continuous processing (see Dietrich, 2001a, and also for some central references). Very briefly, almost all notions of computation are broad enough to include devices seemingly dependent on continuity, like thermostats and mousetraps. In short, there is no robust distinction between analog and digital computation (or processing). The distinction is mostly in the mind of the observer. Again, see Dietrich (2001a). We also refer the reader back to the First AI War, on computation. Finally, there are notions of computation defined over the reals, which is a continuous set.

Sometimes, such machines are stronger than Turing machines. Alas, they are also non-physical. See Blum, Shub, and Smale, 1989.

8. The meaningless processing view is widely held, but it is far from obvious that it is correct: computers may be fully semantically quite naturally. See Dietrich, 1989. And see the discussion of Rapaport's *syntactic semantics* in the interlude, below.

9. The use of pencil and paper was how Searle originally presented the argument. If you want, you can update the method to sending and receiving texts you do not understand together with a database of rules and instructions telling you what to write down given the texts you receive.

10. This debate between Dretske and Rapaport extends deeply into computer science and has strong echoes in the present. We discuss these in Third War Supplement 5.3: Semantics and Causation.

11. Other AI researchers opted to locate the problem of unintelligent machines in other deficits. An important example: Douglas Lenat suggested that the real lack in machines was *common sense* and an *inability to cope with surprises*. The required fix was to build a computer with a vast, encyclopedic amount of knowledge. This resulted in one of the most controversial and fascinating projects in the history of AI: the Cyc Project. See, for example, Lenat, Prakash, and Shepherd (1986).

12. See Fodor, 1980, for the requirement that thinking has two separate dimensions – the content of the thought (e.g. pencils versus pens) and the type of thought (e.g. beliefs versus intentions).

13. In his 1979 paper, McCarthy argued explicitly that it is legitimate to ascribe beliefs to thermostats. However, he also explicitly argued that such legitimacy is governed by *pragmatism*: ascribing beliefs to thermostats is legitimate when it is useful. McCarthy went on to suggest that ascribing beliefs to people is also governed by pragmatism: we ascribe beliefs to people because it is useful to do so. So, it looks like McCarthy is at best an agnostic when it comes to the question 'Do thermostats (and people) *really* have beliefs, in a way beyond pragmatics?' The complexity of, and necessary interrelations between, desiring, believing, intending, planning, and reasoning, etc. are well-presented in Bratman, 1987.

14. For a time in the 1960s and 1970s, the functionalist approach was *the* central theory of the mind adopted by philosophers. This approach was introduced by Hilary Putnam in 1960 in his famous paper 'Minds and Machines'. Functionalism led naturally to computationalism, the theory that the brain is a kind of computer and the mind is its working program. See Dietrich, 1990. See also Fodor, 1980. But functionalism failed to explain the mind in any interesting detail. See e.g. Putnam, 1992.

15. Newell and Simon (1976) say: 'A physical symbol system consists of a set of entities, called symbols, which are physical patterns that can occur as

components of another type of entity called an expression (or symbol structure). Thus, a symbol structure is composed of a number of instances (or tokens) of symbols related in some physical way (such as one token being next to another). At any instant of time the system will contain a collection of these symbol structures. Besides these structures, the system also contains a collection of processes that operate on expressions to produce other expressions: processes of creation, modification, reproduction and destruction. *A physical symbol system is a machine that produces through time an evolving collection of symbol structures . . .*' p. 116 (our emphasis). So it is clear that Newell and Simon focus only on the symbol aspect of thinking. Their definition cannot distinguish between an ordinary computer and a human. But ours can, as the robot sharpener shows. And as should be obvious now, the distinction between an ordinary computer and a human is vast.

16. Dennett is one philosopher who has embraced wholeheartedly the pragmatic idea that aboutness is mere doing: he doesn't require that the doing be robust, like the kind of doing performed by the robot sharpener. Dennett ascribes aboutness to simple chess playing computers. See his: 1978, esp. ch 1; and his 1983, 1987, and 1988. Clearly Dennett is a friend to AI.

17. For the order-level of phenomenal judgments, see Chalmers, 1996, pp. 175–6. Chalmers also says there are third-order judgments about types of conscious experiences. It is not clear if there is a fourth level, and if there is not, why not. But clearly there cannot be an infinite number of levels – the mind is finite.

18. In philosophy, this may be changing, slowly. Some philosophers have started exploring the relation between aboutness and consciousness. See Kriegel, 2013. Some philosophers phrase the relation between consciousness and aboutness by saying that the former *grounds* the latter, see Pautz, 2013. The trouble is that one now needs a theory of grounding. Currently agreement on such a theory remains elusive. For one example, see Fine, 2012.

19. See https://plato.stanford.edu/entries/abduction.

20. See, for example, Drake, 2016, and to get started, see https://en.wikipedia.org/wiki/Moon_illusion.

21. The theory presented here about knowledge requiring consciousness has similarities to some ideas in Chalmers's work. See, esp., his 2003.

Chapter 6

1. A phylum is a taxonomic grouping of organisms, just below the grouping Kingdom, that focuses only on the overall body plan. Humans, along with

the other mammals, as well as birds, reptiles, fish, and amphibians, are in the phylum Chordata: animals with a backbone or dorsal nerve chord.

2. See Dietrich, 2011, for an explanation of why being revealed as a philosophy problem makes the frame problem, in particular, and any problem, in general, intractable. The short form is that philosophy can make no progress on its problems; it can merely restate the problems in newer vocabularies. Also, and more concretely, Dietrich and Fields, in their 2020 paper, show that the frame problem and the halting problem are in fact equivalent. So, the frame problem is provably undecidable. This means that we can at best only solve well-behaved, small versions of the frame problem. Below, we discuss this result further.

3. Mechanism has an ancient history. In 1902, an archeologist discovered what is called today the *Antikythera Mechanism*. This was a Greek analogue computer used for calculating astronomical events decades in the future. The date of the building of the mechanism is somewhat uncertain, but experts locate its creation between 205 BCE and 70–60 BCE.

4. Logic's central notion of *validity* also doesn't deal with change. Validity is the property of an argument where the conclusion must be true if the premises are. This is a wonderful property. The rule of inference, *modus ponens*, provides a clear example. Here's the rule: 'If P implies Q, and P is true, then Q is true'. The two premises of modus ponens are (1) P implies Q, and (2) P. Q must be true, if the two premises are. But again, change is not handled at all. All modus ponens tells us is that *If* the two premises are true, *then* Q must also be true. All of this is static. Change is going on (or can be), but classical logic ignores it.

5. We are ignoring all odd happenings like the pencil gets sharpened until it is only pencil wood shavings and graphite dust, leaving only the eraser.

6. About the second (and third) way McCarthy and Hayes say: 'Another approach to the frame problem *may* follow from the methods of the next section [sec 4.4]; and in part 5 we mention a third approach which *may* be useful, although we have not investigated it at all fully' (our emphases).

7. Compare to the third to last paragraph in Matt Paisner's 'Logicist AI in McCarthy and Hayes (1969)' at https://www.cs.umd.edu/class/fall2012/cmsc828d/oldreportfiles/paisner1.pdf.

8. The scientific literature on analogy-making is large. One might start with the books Gentner et. al, 2001, Holyoak and Thagard, 1995, and Nersessian, 2008.

Chapter 7

1. See https://en.wikipedia.org/wiki/Modal_logic for an introduction.
2. Here is what this proof says in English. Suppose P is some claim, say that having a big brain explains consciousness. Step 1 (Premise): In virtue of just understanding P, one can see that it is possibly not true (there seem to be conscious animals with small brains, e.g. lobsters). Step 2: Therefore P is not necessarily true (this is a principle of modal logic: any claim that is possibly not true is automatically not necessarily true, by the definitions of 'possibly' and 'necessarily'). Step 3 (Premise): If P is true at all it must be necessarily true (and this is logically equivalent to 'if P is not necessarily true, then it is not true' which is what Step 3 says). Therefore P is not true by modus ponens ('if A then B; A; therefore B'). So, having a big brain does not explain consciousness.

Chapter 8

1. See a recording of these comments on CNBC: https://www.cnbc.com/2017/11/06/stephen-hawking-ai-could-be-worst-event-in-civilization.html.
2. Much of this section is adapted and updated from: Sullins, J. P. (2013). 'An Ethical Analysis of the Case for Robotic Weapons Arms Control', in the proceedings for the Fifth International Conference on Cyber Conflict, K. Podins, J. Stinissen, M. Maybaum (Eds.), NATO CCD COE Publications, Tallinn, Estonia.
3. There are many ethical issues regarding the impact of these technologies, too many to cover in detail here. But there are good reports that the interested reader can access such as Lin, et. al, *Autonomous military robotics: risk, ethics, and design*, and Christen et al. *An Evaluation Schema for the Ethical Use of Autonomous Robotic Systems in Security Application*. Lin et al., available at http://ethics.calpoly.edu/ONR_report.pdf and Christen et al., available at SSRN https://papers.ssrn.com/sol3/papers.cfm?abstract_id=3063617.
4. Of course, if the machine is conscious, then it may have the right to self-preservation. Once again, this shows that we are forced to make decisions in the face of extremely important missing information, and as argued in Chapters 5 and 7, getting this information in the case of consciousness looks to be impossible.

Bibliography

Altman, J. (2009). Preventive arms control for uninhabited military vehicles. In: Capurro, R. and Nagenborg, M. (Eds.) *Ethics and Robotics*. Heidelberg: AKA Verlag, pp. 69–82.

Anderson, M. L. (2003). Embodied cognition: A field guide. *Artificial Intelligence* 149, 91–130.

Anderson, M. and Anderson, S. (2011). *Machine Ethics*. Cambridge: Cambridge University Press.

Aristotle (1985). *Nichomachean Ethics*. (Irwin, T., Ed.) Indianapolis, IN: Hackett.

Arkin, R. C. (November 2007). Governing Lethal Behavior: Embedding Ethics in a Hybrid Deliberative/Reactive Robot Architecture, Technical Report GIT-GVU-07-11, Mobile Robot Laboratory, College of Computing, Georgia Institute of Technology. http://www.cc.gatech.edu/ai/robot-lab/online-publications/formalizationv35.pdf (accessed 25 June 2019).

Arkin, R. C. (2010). The case for ethical autonomy in unmanned systems. *Journal of Military Ethics* 9(4), 332–41.

Arquilla, J. (2010). The New Rules of War, Foreign Policy, March/April. http://www.foreignpolicy.com/articles/2010/02/22/the_new_rules_of_war (accessed 25 January 2013).

Arquilla, J. (2012). Cyberwar is already upon us: But can it be controlled? *Foreign Policy*, March/April. http://www.foreignpolicy.com/articles/2012/02/27/cyberwar_is_already_upon_us (accessed 25 January 2013).

Asaro, P. M. (2008). How just could a robot war be? In: Brey, P., Briggle, A. and Waelbers, K. (Eds.) *Current Issues in Computing and Philosophy*. Amsterdam: Ios Press, pp. 50–64.

Asaro, P. M. (2011). Military robots and just war theory. In: Dabringer, G. (Ed.) *Ethical and Legal Aspects of Unmanned Systems*. Vienna: Institut für Religion und Frieden, pp. 103–19.

Asimov, I. (1950). *I, Robot*. New York: Gnome Press.

Baars, B. J. (2005). Global workspace theory of consciousness: Toward a cognitive neuroscience of human experience. *Progress in Brain Research* 150, 45–53.

Baars, B. J., Franklin, S. and Ramsoy, T. Z. (2013). Global workspace dynamics: Cortical 'binding and propagation' enables conscious contents. *Frontiers in Psychology* 4, 200.

Ball, G., Alijabar, P., Zebari, S., Tusor, N., Arichi, T., Merchant, N., Robinson, E. C., Ogundipe, E., Ruekert, D., Edwards, A. D. and Counsell, S. J. (2014). Rich-club organization of the newborn human brain. *Proceedings of the National Academy of Sciences USA* 111, 7456–61.

Baluška, F. and Levin, M. (2016). On having no head: Cognition throughout biological systems. *Frontiers in Psychology* 7, 902.

Bargh, J. A. and Ferguson, M. J. (2000). Beyond behaviorism: On the automaticity of higher mental processes. *Psychological Bulletin* 126(6), 925–45.

Bargh, J. A., Schwader, K. L., Hailey, S. E., Dyer, R. L. and Boothby, E. J. (2012). Automaticity in social-cognitive processes. *Trends in Cognitive Sciences* 16(12), 593–605.

Bartneck, C., Yogeeswaran, K., Ser, Q. M., Woodward, G., Wang, S., Sparrow, R. and Eyssel, F. (2018). Robots and racism. *Proceedings of 2018 ACM/IEEE International Conference on Human-Robot Interaction (HRI 18)* ACM, pp. 196–204.

Barwise, J. and Perry, J. (1983). *Situations and Attitudes.* Cambridge, MA: Bradford Books, MIT Press.

Barwise, J. and Seligman, J. (1997). *Information Flow: The Logic of Distributed Systems. Cambridge Tracts in Theoretical Computer Science 44.* Cambridge, UK: Cambridge University Press.

Beall, J., Brady, R., Dunn, J. M. et al. (2012). On the Ternary Relation and Conditionality, *Journal of Philosophical Logic* 41, 595–612.

Bechtel, W. (1998). Representations and cognitive explanations: Assessing the dynamicist's challenge in cognitive science. *Cognitive Science* 22(3), 295–318.

Bechtel, W. and Abrahamsen, A. (1991). *Connectionism and the Mind: An Introduction to Parallel Processing in Networks.* Cambridge, MA: Blackwell.

Bellmund, J. et al. (2018). Navigating cognition: Spatial codes for human thinking, *Science* 362, eaat6766.

Bello, P. and Bringsjord, S. (2013). On how to build a moral machine. *Topoi* 32(2), 251–66.

Bentham, J. (1789). *An Introduction to the Principles of Morals and Legislation.* In: *The Collected Works of Jeremy Bentham* (1970), Burns, J. H. and Hart, H. L. A. (Eds.) Oxford: Oxford University Press.

Bischoff, P. and Rundshagen, I. (2007). Awareness during general anesthesia. *Deutsches Ärzteblatt International* 108(1–2), 1–7.

Block, N. (1995). On a confusion about a function of consciousness. *Behavioral and Brain Sciences* 18(2), 227–47.

Blum, L. Shub, M. and Smale, S. (1989). On a theory of computation and complexity over the real numbers: NP-completeness, recursive functions, and universal machines. *Bulletin of the American Mathematical Society* 21(1) (July), 1–46.

Boly, M., Sanders, R. D., Mashour, G. A. and Laureys, S. (2013). Consciousness and responsiveness: Lessons from anaesthesia and the vegetative state. *Current Opinion in Anesthesiology* 26(4), 444–9.

Bonnefon, J.-F., Shariff, A. and Rahwan, I. (2016). The social dilemma of autonomous vehicles. *Science* 352, 1573–6.

Boole, G. (1854). *The Laws of Thought.* London: Macmillan.

Bostrom, N. (2014). *Superintelligence: Paths, Dangers, Strategies.* Oxford, UK: Oxford University Press.

Brachman, R. and Smith, B., Eds. (1980). Special Issue on Knowledge Representation, *SIGART Newsletter* (Special Interest Group on Artificial Intelligence; a former publication of the Association of Computing Machinery) https://dl.acm.org/citation.cfm?id=1056752 (accessed 7 July 2019).

Bratman, M. (1987). *Intention, Plans, and Practical Reason.* Cambridge, MA: Harvard University Press.

Brentano, F. (1874). *Psychology from an Empirical Standpoint.* London: Routledge and Kegan Paul.

Brewer, J. A., Worhunsky, P. D., Gray, J. R., Tang, Y.-Y., Weber, J. and Kober, H. (2011). Meditation experience is associated with differences in default mode network activity and connectivity. *Proceedings of the National Academy of Sciences USA* 108, 20254–9.

Bringsjord, S. and Taylor, J. (2012). The divine-command approach to robot ethics. In: Lin, P. Abney, K. and Bekey, G. (Eds.) *Robot Ethics: The Ethical and Social Implications of Robotics.* Cambridge, MA: MIT Press, pp. 85–108.

Brooks, R. (1989). A robot that walks; Emergent behaviors from a carefully evolved network. *Neural Computation* 1, 253–62.

Brooks, R. (1991). Intelligence without representation. *Artificial Intelligence* 47, 139–59.

Brooks, R. (1999). *Cambrian Intelligence: The Early History of the New AI.* Cambridge, MA: MIT Press.

Buckner, R., Andrews-Hanna, J. and Schacter, D. (2008). The brain's default network: Anatomy, function, and relevance to disease. *Annals of the New York Academy of Sciences* 1124, 1–38.

Buzsáki, G. and Llinás, R. (2017). Space and time in the brain. *Science* 358, 478–82.

Cangelosi, A. and Schlesinger, M. (2015). *Developmental Robotics: From Babies to Robots.* Cambridge, MA: MIT Press.

Carhart-Harris, R. et al. (2012). Neural correlates of the psychedelic state as determined by fMRI studies with psilocybin. *Proceedings of the National Academy of Sciences USA* 109(6), 2138–43.

Carhart-Harris, R. et al. (2016). Neural correlates of the LSD experience revealed by multimodal neuroimaging. *Proceedings of the National Academy of Sciences USA* 113(17), 4853–8.

Carlén, M. (2017). What constitutes the prefrontal cortex? *Science* 358, 478–82.

Carpenter, J., Davis, J. M., Erwin-Stewart, N., Lee, T. R., Bransford, J. D. and Vye, N. (2009). Gender representation and humanoid robots designed for domestic use. *International Journal of Social Robotics* 1, 261.

Carter, O., Hohwy, J., van Boxtel, J., Lamme, V., Block, N., Koch, C. and Tsuchiya, N. (2018). Conscious machines: Defining questions. *Science* 359, 400.

Cerullo, M. A. (2015). The problem with Phi: A critique of Integrated Information Theory. *PLoS Computational Biology* 11, e1004286.

Chalmers, D. J. (1995). Minds, machines, and mathematics. *Psyche* 2(1), 11–20.

Chalmers, D. J. (1996). *The Conscious Mind: In Search of a Fundamental Theory*. Oxford: Oxford University Press.

Chalmers, D. (2003). The content and epistemology of phenomenal belief. In: Smith, Q. and Jokic, A. (Eds.) *Consciousness: New Philosophical Perspectives*. Oxford: Oxford University Press, pp. 220–72.

Chalmers, D. (2010). The Singularity: A philosophical analysis. *Journal of Consciousness Studies* 17, 7–65. See also http://consc.net/papers/singularity. pdf.

Chater, N. (2018). *The Mind Is Flat*. London: Allen Lane.

Chemero, A. (2009). *Radical Embodied Cognition*. Cambridge, MA: MIT Press.

Chomsky, N. (1957). *Syntactic Structures*. The Hague/Paris: Mouton.

Chopra, S. (2011). Taking the moral stance: Morality, robots, and the intentional stance. In: Van Den Berg, B. and Klaming, L. (Eds.) *Technologies on the Stand: Legal and Ethical Questions in Neuroscience and Robotics*. Nijmegen: Wolf Legal Publishers, pp. 285–95.

Chrisley, R. (2003). Embodied artificial intelligence. *Artificial Intelligence* 149, 131–50.

Christian, M., Burri, T., Chapa, J. O., Salvi, R., Santoni de Sio, F. and Sullins, J. (2017). An Evaluation Schema for the Ethical Use of Autonomous Robotic Systems in Security Applications (1 November 2017). https://papers.ssrn. com/sol3/papers.cfm?abstract_id=3063617 (accessed 24 June 2019).

Chudnoff, E. (2013). *Intuition*. Oxford, UK: Oxford University Press.

Churchland, P. (1989). *A Neurocomputational Perspective: The Nature of Mind and the Structure of Science*. Cambridge, MA: MIT Press.

Churchland, P. (1995). *The Engine of Reason, The Seat of the Soul: A Philosophical Journey into the Brain*. Cambridge, MA: MIT Press.

Churchland, P. (1996). *Neurophilosophy: Toward a Unified Science of the Mind-Brain*. Cambridge, MA: MIT Press.

Clark, A. (1993). *Associative Engines: Connectionism, Concepts, and Representational Change*. Cambridge, MA: MIT Press.

Clark, A. (2003). *Natural Born Cyborgs: Minds, Technologies, and the Future of Human Intelligence*, Oxford, UK: Oxford University Press.

Clark, A. (2008). *Supersizing the Mind: Embodiment, Action, and Cognitive Extension*. Oxford, UK: Oxford University Press.

Clark, A. (2013). Whatever next? Predictive brains, situated agents, and the future of cognitive science. *Behavioral and Brain Sciences* 36, 181–204.

Clark, A. (2017). How to knit your own Markov blanket. In: Metzinger, T. and Wiese, W. (Eds.). *Philosophy and Predictive Processing: 3*. Frankfurt am Main: MIND Group.

Coeckelbergh, M. (2012). Can we trust robots? *Ethics and Information Technology* 14, 53–60.

Cooper, S. and van Leeuwen, J. (Eds.) (2013). *Alan Turing: His Work and Impact*. Amsterdam: Elsevier.

Copeland, J. (1993). *Artificial Intelligence: A Philosophical Introduction*. Oxford, UK: Blackwell.

Craig, A. D. (2002). How do you feel? Interoception: The sense of the physiological condition of the body. *Nature Reviews Neuroscience* 3, 655–66.

Craig, A. D. (2009). How do you feel–now? The anterior insula and human awareness. *Nature Reviews Neuroscience* 10, 59–70.

Craig, A. D. (2010). The sentient self. *Brain Structure and Function* 214, 563–77.

Csikszentmihályi, M. (1990). *Flow: The Psychology of Optimal Experience*. New York: Harper and Row.

Dabringer, G., Ed. (2011). *Ethical and Legal Aspects of Unmanned Systems*. Vienna: Institut für Religion und Frieden.

Damasio, A. (1994). *Descartes' Error: Emotion, Reason, and the Human Brain*, New York: Quill.

Damasio, A. R. (1999). *The Feeling of What Happens*. New York: Harcourt.

Damasio, A. (2003). *Looking for Spinoza: Joy, Sorrow, and the Feeling Brain*. Orlando: Harcourt.

De Jaegher, H. and Di Paolo, E. (2007). Participatory sense-making: An enactive approach to social cognition. *Phenomenology and the Cognitive Sciences* 6(4), 485–507.

Debruyne, H., Portzky, M., Van den Eynde, F. and Audenaert, K. (2009). Cotard's syndrome: A review. *Current Psychiatry Reports* 11(3), 197–202.

Dehaene, S. and Naccache, L. (2001). Towards a cognitive neuroscience of consciousness: Basic evidence and a workspace framework. *Cognition* 79, 1–37.

Dehaene, S., Charles, L., King, J.-R. and Marti, S. (2014). Toward a computational theory of conscious processing. *Current Opinion in Neurobiology* 25, 76–84.

Dehaene, S., Lau, H. and Kouider, S. (2017). What is Consciousness, and Could Machines Have It? *Science* 358, 486–92.

Dennett, D. (1978). *Brainstorms: Philosophical essays on mind and psychology*. Cambridge, MA: MIT Press/Bradford Books.

Dennett, D. (1983). Intentional systems in cognitive ethology: The 'Panglossian paradigm' defended. *Behavioral and Brain Sciences* 6, 343–90. (Reprinted as Chapter 7 of Dennett, *The Intentional Stance.*)

Dennett, D. (1987a). *The Intentional Stance.* Cambridge, MA: MIT Press/ Bradford Books.

Dennett, D. (1987b). Cognitive wheels: The frame problem in AI. In: Pylyshyn, Z. (Ed.) *The Robot's Dilemma: The Frame Problem in Artificial Intelligence.* Norwood, NJ: Ablex, pp. 41–64.

Dennett, D. (1988). Précis of *The Intentional Stance. Behavioral and Brain Sciences* 11, 495–546.

Dennett, D. (1991). *Consciousness Explained.* New York: Little, Brown, and Co.

Dennett, D. C. (2017). *From Bacteria to Bach and Back: The Evolution of Minds.* London: Penguin.

Deutsch, D. (1985). Quantum theory, the Church-Turing principle and the universal quantum computer. *Proceedings of the Royal Society of London A* 400, 97–117.

Dewey, J. (1928/1984). Body and mind. In: Boydston, J. A. and Baysinger, P. (Eds.) *John Dewey: The Later Works, 1925–1953. Volume 3: 1927–1928.* Carbondale, IL: Southern Illinois University Press, pp. 25–40.

Dewey, J. (1998a). The influence of Darwinism on philosophy. In: Hickman, L. and Alexander, T. (Eds.) *The Essential Dewey, Volume 1.* Bloomington, IN: Indiana University Press, pp. 39–45.

Dewey, J. (1998b). Evolution and ethics. In: Hickman, L. and Alexander, T. (Eds.). *The Essential Dewey, Volume 2.* Bloomington, IN: Indiana University Press, pp. 225–36.

Dietrich, E. (1989). Semantics and the computational paradigm in cognitive psychology. *Synthese* 79, pp. 119–41.

Dietrich, E. (1990). Computationalism. *Social Epistemology* 4(2), pp. 135–54. (with commentary).

Dietrich, E. (1995). AI and the mechanistic forces of darkness, *Journal of Experimental and Theoretical Artificial Intelligence* 7(2), pp. 155–61.

Dietrich, E. (2000). Analogy and conceptual change, or You can't step into the same mind twice. In: Dietrich, E. and Markman, A. (Eds.) *Cognitive Dynamics: Conceptual Change in Humans and Machines.* Mahwah, NJ: Lawrence Erlbaum, pp. 265–94.

Dietrich, E. (2001a). The Ubiquity of Computation. https://binghamton. academia.edu/EricDietrich/Papers.

Dietrich, E. (2001b). It does so! Review of Jerry Fodor's *The Mind Doesn't Work That Way: The Scope and Limits of Computational Psychology. AI Magazine* 22(4), 141–4.

Dietrich, E. (2010). Analogical insight: toward unifying categorization and analogy. *Cognitive Processing* 11(4), pp. 331–45.

Dietrich, E. (2011a). There is no progress in philosophy. *Essays in Philosophy* 12(2), pp. 329–44.

Dietrich, E. (2011b). *Homo sapiens* 2.0 Why we should build the better robots of our nature. In: Anderson, M. and Anderson, S. (Eds.) *Machine Ethics.* Cambridge, UK: Cambridge University Press, pp. 531–8.

Dietrich, E. (2014). Are You a Chinese Room?: The Fantastic Voyage Variation. Course manuscript, Binghamton University, New York. Available at http://bingweb.binghamton.edu/~dietrich/Selected_Publications.html and https://www.researchgate.net/profile/Eric_Dietrich/research.

Dietrich, E. and Fields, C. (1996). The role of the frame problem in Fodor's modularity thesis: A case study of rationalist cognitive science. In: Ford, K. and Pylyshyn, Z. (Eds.) *The Robot's Dilemma Revisited.* Norwood, NJ: Ablex. pp. 9–24.

Dietrich, E. and Fields, C. (2020). Equivalence of the frame and halting problems. *Algorithms* 13, 175.

Dietrich, E. and Gillies, A. (2001). Consciousness and the limits of our imaginations. *Synthese* 126(3), pp. 361–81.

Dietrich, E. and Hardcastle, V. G. (2004). *Sisyphus's Boulder: Consciousness and the Limits of the Knowable.* Amsterdam: John Benjamins.

Dietrich, E. and Markman, A. B. (2001). Dynamical description versus dynamical modeling: Reply to Chemero. *Trends in Cognitive Sciences* 5(8), 332.

Dietrich, E. and Markman, A. B. (2003). Discrete Thoughts: Why cognition must use discrete representations. *Mind and Language* 18(1), 95–119.

Drake, N. (2016). Why the moon looks bigger near the horizon. *National Geographic.* https://news.nationalgeographic.com/2016/12/moon-illusion-explained-horizon-size-supermoon-space-science (accessed 24 June 2019).

Dretske, F. (1985). Machines and the mental, *Proceedings and Addresses of the American Philosophical Association* 59(1), 23–33.

Dreyfus, H. (1972). *What Computers Can't Do.* Cambridge, MA: MIT Press.

Dyer, M. G. (1990). Intentionality and computationalism: Minds, machines, Searle and Harnad. *Journal of Experimental and Theoretical Artificial Intelligence* 2(4), 303–91.

Edelman, G. (1987). *Neural Darwinism. The Theory of Neuronal Group Selection.* Basic Books.

Eichenbaum, H., Yonelinas, A. R. and Ranganath, C. (2007). The medial temporal lobe and recognition memory. *Annual Review of Neuroscience* 30, 123–52.

Ess, C. (2009). *Digital Media Ethics.* Cambridge, UK: Polity Press.

Evans, J. St. B. T. (2003). In two minds: Dual process accounts of reasoning. *Trends in Cognitive Sciences* 7, pp. 454–9.

Evans, J. St. B. T. and Stanovich, K. E. (2013). Dual-process theories of higher cognition: Advancing the debate. *Perspectives on Psychological Science* 8(3), pp. 223–41.

Eyssel, F. and Hegel, F. (2012). (S)he's got the look: Gender stereotyping of robots. *Journal of Applied Social Psychology* 42(9), pp. 2213–30.

Eyssel, F. and Kuchenbrandt, D. (2012). Social categorization of social robots: Anthropomorphism as a function of robot group membership. *British Journal of Social Psychology* 51, pp. 724–31.

Eyssel, F., Kuchenbrandt, D., Bobinger, S., de Ruiter, L. and Hegel, F. (2012). If you sound like me, you must be more human: On the interplay of robot and user features on human-robot acceptance and anthropomorphism. *Proceedings of 7th ACM/IEEE Conference on HRI*, pp. 125–6.

Fama, R., Pitel, A.-L. and Sullivan, E. V. (2012). Anterograde episodic memory in Korsakoff Syndrome. *Neuropsychological Review* 22(2), 93–104.

Feigenbaum, E. A. and Feldman, J., Eds. (1963). *Computers and Thought*. New York: McGraw-Hill.

Fesmire, S. (2003). *John Dewey and Moral Imagination: Pragmatism in Ethics*. Bloomington, IN: Indiana University Press.

Feynman, R. P. (1982). Simulating physics with computers. *International Journal of Theoretical Physics* 21, 467–88.

Fields, C. (1989). Consequences of nonclassical measurement for the algorithmic description of continuous dynamical systems. *Journal of Experimental and Theoretical Artificial Intelligence* 1, 171–8.

Fields, C. (2012). The very same thing: Extending the object token concept to incorporate causal constraints on individual identity. *Advances in Cognitive Psychology* 8, 234–47.

Fields, C. (2013). How humans solve the Frame problem. *Journal of Experimental and Theoretical Artificial Intelligence* 25, 441–56.

Fields, C. (2014). Equivalence of the symbol grounding and quantum system identification problems. *Information* 5, 172–89.

Fields, C. (2016). Visual re-identification of individual objects: A core problem for organisms and AI. *Cognitive Processing* 17(1), 1–13.

Fields, C., Hoffman, D. D., Prakash, C. and Prentner, R. (2017). Eigenforms, interfaces and holographic encoding: Toward an evolutionary account of objects and spacetime. *Constructivist Foundations* 12, 265–91.

Fields, C., Hoffman, D. D., Prakash, C. and Singh, M. (2018). Conscious agent networks: Formal analysis and application to cognition. *Cognitive Systems Research* 47, 186–213.

Fine, K. (2012). Guide to ground. In: Correia, F. and Schnieder, B. (Eds.) *Metaphysical Grounding*. Cambridge, UK: Cambridge University Press, pp. 37–80.

Firestone, C. and Scholl, B. J. (2016). Cognition does not affect perception: Evaluating the evidence for 'top-down' effects. *Behavioral and Brain Sciences* 39, e229.

Floridi, L. and Sanders, J. W. (2004). On the morality of artificial agents. *Minds and Machines* 14(3), 349–79.

Fodor, J. (1975). *The Language of Thought*. Cambridge, MA: Harvard University Press.

Fodor, J. A. (1980). Methodological solipsism considered as a research strategy in cognitive psychology. *Behavioral and Brain Sciences* 3 (1), 63–73.

Fodor, J. A. (1983). *The Modularity of Mind*. Cambridge, MA: MIT Press.

Fodor, J. A. (1985). Précis of *The Modularity of Mind. Behavioral and Brain Sciences* 8, 1–42.

Fodor, J. A. (1986). Why paramecia don't have mental representations. *Midwest Studies in the Philosophy of Science* 10, 3–23.

Fodor, J. A. (1987). Modules, frames, fridgeons, sleeping dogs, and the music of the spheres. In: Pylyshyn, Z. (Ed.) *The Robot's Dilemma: The Frame Problem in Artificial Intelligence*. Norwood, NJ: Ablex, pp. 139–49.

Fodor, J. A. (1998). *In Critical Condition*. Cambridge, MA: MIT Press.

Fodor, J. A. (2000). *The Mind Doesn't Work That Way: The Scope and Limits of Computational Psychology*. Cambridge, MA: MIT Press.

Fodor, J. A. and McLaughlin, B. P. (1990). Connectionism and the problem of systematicity: Why Smolensky's solution doesn't work. *Cognition* 35, 183–204.

Fodor, J. A. and Pylyshyn, Z. (1988). Connectionism and cognitive architecture: A critical analysis. *Cognition* 28, 3–71.

Foot, Philippa (1967). The problem of abortion and the doctrine of the double effect. *Oxford Review* 5, 5–15.

Forsythe, D. (2001). *Studying Those Who Study Us: An Anthropologist in the World of Artificial Intelligence*. Stanford, CA: Stanford University Press.

Frankish, K. (2016). Illusionism as a theory of consciousness. *Journal of Consciousness Studies* 23(11–12), 11–39.

Franklin, S., Madl, T., D'Mello, S. and Snaider, J. (2014). LIDA: a systems-level architecture for cognition, emotion and learning. *IEEE Transactions on Autonomous Mental Development* 6, 19–41.

Frege, G. (1879). *Begriffsschrift: A formal language, modeled upon that of arithmetic, for pure thought*. Halle: Verlag von Louis Nebert.

Friston, K. J. (2010). The free-energy principle: A unified brain theory? *Nature Reviews Neuroscience* 11, 127–38.

Froese, T. and Ziemke, T. (2009). Enactive artificial intelligence: Investigating the systemic organization of life and mind. *Artificial Intelligence* 173, 466–500.

Gabriel, G., Hermes, H., Kambartel, F., Thiel, C., Veraart, A., McGuinness, A. and Kaal, H., Eds. (1980). *Philosophical and Mathematical Correspondence*. Oxford: Blackwell Publishers.

Gagliano, M., Renton, M., Depczynski, M. and Mancuso, S. (2014). Experience teaches plants to learn faster and forget slower in environments where it matters. *Oecologia* 175(1), 63–72.

Gallese, V. and Lakoff, G. (2005). The brain's concepts: The role of the sensory-motor system in conceptual knowledge. *Cognitive Neuropsychology* 22(3), 455–79.

Gallese, V. and Sinigaglia, C. (2011). What's so special about embodied simulation? *Trends in Cognitive Sciences* 15(11), 512–19.

Gandy, R. (1996). Human versus machine intelligence. In: Millican, P. and Clark, A. (Eds.) *Machines and Thought: The Legacy of Alan Turing*, Volume 1. Oxford, UK: Oxford University Press. pp. 125–36.

Gao, T., McCarthy, G. and Scholl, B. J. (2010). The wolfpack effect: Perception of animacy irresistibly influences interactive behavior. *Psychological Science* 21, 1845–53.

Gao, W., Alcauter, S., Smith, J. K., Gilmore, J. H. and Lin, W. (2015). Development of human brain cortical network architecture during infancy. *Brain Structure and Function* 220, 1173–86.

Garson, J. (2018). Connectionism. In: Zalta, E. N. (Ed.) *The Stanford Encyclopedia of Philosophy* (Fall 2018 Edition), https://plato.stanford.edu/archives/fall2018/entries/connectionism (accessed 24 June 2019).

Gathmann, F., Gebauer, M., Medick, V. and Weiland, S. (2013). Deutschlands Drohnenpläne: Merkel rüstet auf, *Spiegel Online*, 25 January. http://www.spiegel.de/politik/deutschland/kampfdrohnen-plaene-der-regierung-stossen-auf-heftigen-widerstand-a-879701.html (accessed 6 February 2013).

Gentner, D. (2010). Bootstrapping the mind: Analogical processes and symbol systems. *Cognitive Science* 34, 752–75.

Gentner, D., Holyoak, K. and Kokinov, B. (2001). *The Analogical Mind: Perspectives from Cognitive Science*. Cambridge, MA: MIT Press.

Gerdes, J. and Thornton, S. (2015). Implementable ethics for autonomous vehicles. In: Maurer, M., Gerdes, J., Lenz, B. and Winner, H. (Eds.) *Autonomes Fahren*. Berlin: Springer, pp. 87–102.

Gibson, J. J. (1979). *The Ecological Approach to Visual Perception*. Boston: Houghton Mifflin.

Gidon, A., Zolnik, T. A., Fidzinski, P., Bolduan, F., Papoutsi, A., Poirazi, P., Holtkamp, M. Vida, I. and Larkum, M. E. (2020). Dendritic action potentials and computation in human layer 2/3 cortical neurons. *Science* 367, 83–7.

Gödel, K. (1931). *Über die Vollständigkeit des Logikkalküls*. PhD thesis, University Of Vienna.

Gödel, K. (1931). Über formal unentscheidbare sätze der principia mathematica und verwandter systeme, i. *Monatshefte für Mathematik und Physik* 38(1), 173–98.

Godfrey-Smith, Peter (2016). *Other Minds: The Octopus, the Sea, and the Deep Origins of Consciousness*. New York: Farrar, Straus, and Giroux.

Goertzel, B., Lian, R., Arel, I., de Garis, H. and Chen, S. (2010). A world survey of artificial brain projects, part II: Biologically inspired cognitive architectures. *Neurocomputing* 74, 30–49.

Goldstein, R. (2017). Mattering Matters. *Free Inquiry* [ejournal]. 37, 2. See also https://www.secularhumanism.org/index.php/articles/8609.

Govindarajulu, N. S. and Bringsjord, S. (2015). Ethical regulation of robots must be embedded in their operating systems. In: Trappl, R. (Ed.) *A Construction Manual for Robots' Ethical Systems: Requirements, Methods, Implementations*. Berlin: Springer, pp. 85–99.

Graham, G. (2017). Behaviorism. In: Zalta, E. N. (Ed.) *The Stanford Encyclopedia of Philosophy* (Spring 2017 Edition), https://plato.stanford.edu/archives/spr2017/entries/behaviorism (accessed 24 June 2019).

Griffin, J. D. and Fletcher, P. C. (2017). Predictive processing, source monitoring, and psychosis. *Annual Review of Clinical Psychology* 13, 265–89.

Griffiths, R. R., Richards, W. A., McCann, U. and Jesse, R. (2006). Psilocybin can occasion mystical-type experiences having substantial and sustained personal meaning and spiritual significance. *Psychopharmacology* 187, 268–83.

Grodzinsky, F., Miller, K. and Wolf, M. (2011). Developing artificial agents worthy of trust: 'Would you buy a used car from this artificial agent?' *Ethics and Information Technology* 13(1), 17–27.

Grodzinsky, F.; Miller, K.; and Wolf M. J. (2020). Trust in Artificial Agents. In Simon, J. (Ed.) *The Routledge Handbook of Trust and Philosophy*. New York and London: Routledge

Gschwind, M. and Picard, F. (2016). Ecstatic epileptic seizures: A glimpse into the multiple roles of the insula. *Frontiers in Behavioral Neuroscience* 10, 21.

Gunkel, D. (2019). *Robot Rights*. Cambridge, MA: MIT Press.

Hacker, P. M. S. (2002). Is there anything it is like to be a bat? *Philosophy* 77(2), 157–74.

Haltiwanger, J. (2018). Trump inherited Obama's drone war and he's significantly expanded it in countries where the US is not technically at war, *Business Insider*, 27 November. https://www.businessinsider.com/trump-has-expanded-obamas-drone-war-to-shadow-war-zones-2018-11 (accessed 24 June 2019).

Hameroff, S. and Penrose, R. (2014). Consciousness in the universe: A review of the 'Orch OR' theory. *Physics of Life Reviews* 11(1), 39–78.

Hamilton, J. P., Furman, D. J., Chang, C., Thomason, M. E., Dennis, E. and Gotlib, I. H. (2011). Default-mode and task-positive network activity in Major Depressive Disorder: Implications for adaptive and maladaptive rumination. *Biological Psychiatry* 70(4), 327–33.

Harnad, S. (1982). Consciousness: An afterthought. *Cognition and Brain Theory* 5, 29–47.

Harnad, S. (1990). The symbol grounding problem. *Physica D* 42, 335–46.

Harrington, R. and Loffredo, D. A. (2011). Insight, rumination, and self-reflection as predictors of well-being. *Journal of Psychology* 145(1), 39–57.

Hayes, P. (1987). What the frame problem is and isn't. In: Pylyshyn, Z. (Ed.) *The Robot's Dilemma: The Frame Problem in Artificial Intelligence*. Norwood, NJ: Ablex, pp. 123–37.

Heavey, C. L. and Hurlburt, R. T. (2008). The phenomena of inner experience. *Consciousness and Cognition* 17(3), 798–810.

Henkel, L. A. and Carbuto, M. (2008). Remembering what we did: How source misattributions arise from verbalization, mental imagery, and pictures. In: Kelley, M. R. (Ed.) *Applied Memory*. Hauppauge, NY: Nova Science Publishers, pp. 213–34.

Hirstein, W. and Ramachandran, V. S. (1997). Capgras syndrome: A novel probe for understanding the neural representation of the identity and familiarity of persons. *Philosophical Transactions of the Royal Society of London B* 264, 437–44.

Hoffman, D. D. (2016). The interface theory of perception. *Current Directions in Psychological Science* 25, 157–61.

Hoffman, D. D. and Prakash, C. (2014). Objects of consciousness. *Frontiers in Psychology* 5, 577.

Hoffman, D. D., Singh, M. and Prakash, C. (2015). The interface theory of perception. *Psychonomic Bulletin & Review* 22, 1480–506.

Holyoak, K. and Thagard, P. (1995). *Mental Leaps: Analogy in Creative Thought*. Cambridge, MA: MIT Press.

Huang, H., Shu, N., Mishra, V., Jeon, T., Chalak, L., Wang, Z. J., Rollins, N., Gong, G., Cheng, H., Peng, Y., Dong, Q. and He, Y. (2015). Development of human brain structural networks through infancy and childhood. *Cerebral Cortex* 25, 1389–404.

Hubbard, E. M. and Ramachandran, V. S. (2005). Neurocognitive mechanisms of synesthesia, *Neuron* 48, 509–20.

Hume, David (1739–1740). *A Treatise of Human Nature: Being an Attempt to Introduce the Experimental Method of Reasoning into Moral Subjects*. See: *A Treatise of Human Nature* (Oxford Philosophical Texts), Norton, D. F. and Norton, M. J., Eds. Oxford: Clarendon Press, 2000, 2007.

Hummel, J. E. and Holyoak, K. J. (1997). Distributed representations of structure: A theory of analogical access and mapping. *Psychological Review*, 104(3), 427–66.

Hutson, M. (2018). Basic Instincts: Some say artificial intelligence needs to learn like a child. *Science* 360, 845–7.

Ince, D. C. (Ed.) (1992). *Mechanical Intelligence Vol. I*. Amsterdam: Elsevier.

Johnson, M. (1993). *Moral Imagination: Implications of Cognitive Science for Ethics*, Chicago: University of Chicago Press.

Johnson, M. (1996). How moral psychology changes moral theory. In: May, L., Friedman, M. and Clark, C. (Eds.) *Mind and Morals: Essays on Ethics and Cognitive Science*. Cambridge, MA: MIT Press, pp. 45–68.

Josipovic, Z. (2014). Neural correlates of nondual awareness in meditation. *Annals of the New York Academy of Sciences* 1307, 9–18.

Kahneman, D. (2011). *Thinking, Fast and Slow*. New York: Farrar, Straus and Giroux.

Kim, D. I., Manoach, D. S., Mathalon, D. H., Turner, J. A., Mannell, M., Brown, G. G., Ford, J. M., Gollub, R. L., White, T., Wible, C., Belger, A., Bockholt, H. J., Clark, V. P., Lauriello, J., O'Leary, D., Mueller, B. A., Lim, K. O., Andreasen, N., Potkin, S. G. and Calhoun, V. D. (2009). Dysregulation of working memory and default-mode networks in schizophrenia using independent component analysis, an fBIRN and MCIC study. *Human Brain Mapping* 30(11), 3795–811.

Kim, L. (2013). Germany and drones. *International Herald Tribune*, 5 February. http://latitude.blogs.nytimes.com/2013/02/05/germany-and-drones/?nl=opinion&emc=edit_ty_20130205 (accessed 5 February 2013).

Klein, S. B. (2014). What memory is. *WIREs Cognitive Science* 2014, 1333.

Klein, S. B. and Nichols, S. (2012). Memory and the sense of personal identity. *Mind* 121, 677–702.

Kosslyn, S. M., Thompson, W. L. and Ganis, G. (2006). *The Case for Mental Imagery*. New York: Oxford University Press.

Kriegel, U., Ed. (2013). *Phenomenal Intentionality*. Oxford, UK: Oxford University Press.

Kripke, S. A. (1972). Naming and necessity. In: Davidson, D. and Harman, G. (Eds.) *Semantics of Natural Language*. Dordrecht: Springer, pp. 253–355.

Kuchling, F., Friston, K., Georgiev, G. and Levin, M. (2019). Morphogenesis as Bayesian inference: A variational approach to pattern formation and control in complex biological systems. *Physics of Life Reviews* 33, 88–108.

Kurzweil, Ray (2005). *The Singularity is Near*. New York: Viking Books.

Lakoff, G. and Johnson, M. (1980). *Metaphors We Live By*. Chicago: University of Chicago Press.

Lakoff, G. and Johnson, M. (1999). *Philosophy in the Flesh: The Embodied Mind and its Challenge to Western Thought*. New York: Basic Books.

Landler, Mark (2012). Civilian deaths due to drones are not many, Obama says. *The New York Times*, 30 January. http://www.nytimes.com/2012/01/31/world/middleeast/civilian-deaths-due-to-drones-are-few-obama-says.html?_r=0 (accessed 28 March 2013).

LeCun, Y., Bengio, Y. and Hinton, G. (2015). Deep learning. *Nature* 521, 436–44.

Leibniz, G. (1685/1951). The art of discovery. In: Wiener, P. P. (Ed.) *Leibniz: Selections*. New York: Charles Scribner's Sons, pp. 50–8. See also Kulstad, M. and Carlin, L. (2013). Leibniz's Philosophy of Mind. In: Zalta, E. N. (Ed.)

The Stanford Encyclopedia of Philosophy (Winter 2013 Edition), https://plato. stanford.edu/archives/win2013/entries/leibniz-mind (accessed 24 June 2019).

Lenat, D., Prakash, M. and Shepherd, M. (1986). CYC: Using common sense knowledge to overcome brittleness and knowledge acquisition bottlenecks. *AI Magazine* 6(4), 65–85.

Libet, B. (1985a). Unconscious cerebral initiative and the role of conscious will in voluntary action. *Behavioral and Brain Sciences* 8, 529–66.

Libet, B. (1985b). Subjective antedating of a sensory experience and mind-brain theories: Reply to Honderich (1984). *Journal of Theoretical Biology* 114, 563–70.

Libet, B. (1987). Are the mental experiences of will and self-control significant for the performance of a voluntary act? *Behavioral and Brain Sciences* 10, 783–6.

Libet, B. (1989a). The timing of a subjective experience. *Behavioral and Brain Sciences* 12, 183–5.

Libet, B. (1989b). Conscious subjective experience vs. unconscious mental functions: A theory of the cerebral processes involved. In: Cotterill, R. M. J. (Ed.) *Models of brain function.* Cambridge: Cambridge University Press, pp. 325–39.

Libet, B. (1991). Conscious and neural timings. *Nature* 352, 27.

Lin, P. (2015). Why ethics matters for autonomous cars. In: Maurer, M., Gerdes, J., Lenz, B. and Winner, H. (Eds.) *Autonomes Fahren.* Berlin: Springer Verlag, 69–86.

Lin, P. (2018). Who is at fault in Uber's fatal collision? *IEEE Spectrum*, 22 March. https://spectrum.ieee.org/cars-that-think/transportation/safety/reflecting-on-ubers-fatal-crash (accessed 24 June 2019).

Lin, P., Bekey, G. and Abney, K. (2008). Autonomous military robotics: Risk, ethics, and design. California Polytechnic State University, San Luis Obispo. http://ethics.calpoly.edu/ONR_report.pdf (accessed 8 July 2014).

Linnainmaa, S. (1976). Taylor expansion of the accumulated rounding error. *BIT Numerical Mathematics* 16(2), 146–60.

Lloyd, S. (1996). Universal quantum simulators. *Science* 273, 1073–8.

Lucas, J. R. (1961). Minds, Machines, and Gödel, *Philosophy* XXXVI, 112–27. Reprinted in: Sayre, K. M. and Crosson, F. J., Eds. (1963). *The Modeling of Mind.* Notre Dame, IN: Notre Dame Press, pp. 269–70.

Lyubova, N., Ivaldi, S. and Filliat, D. (2016). From passive to interactive object learning and recognition through self-identification on a humanoid robot. *Autonomous Robots* 40, 33–57.

Machery, E. (2009). *Doing without concepts.* New York: Oxford University Press.

Machery, E. (2010). Précis of *Doing without concepts. Behavioral and Brain Sciences* 33(2–3), 195–206.

Madison, G. (1991). Merleau-Ponty's deconstruction of logocentrism. In: Dillon, M. (Ed.) *Merleau-Ponty Vivant.* Albany, SUNY Press, pp. 117–52.

Margolis, E. and Laurence, S. (1999). *Concepts: Core Readings.* Cambridge, MA: MIT Press.

Margolis, E. and Laurence, S. (2014). Concepts. In: Zalta, E. N. (Ed.) *The Stanford Encyclopedia of Philosophy* (Spring 2014 Edition), https://plato. stanford.edu/archives/spr2014/entries/concepts (accessed 24 June 2019).

Mariscal, C. and Doolittle, F. W. (2020). Life and life only: A radical alternative to life definitionism. *Synthese* 197, 2975–2989.

Markman, A. and Dietrich, E. (2000). In defense of representation. *Cognitive Psychology* 40, 138–71.

Masters, R. E. L. and Houston, J. (1966). *The Varieties of Psychedelic Experience.* New York: Holt, Rinehart and Winston.

May, L., Friedman, M. and Clark, C., Eds. (1996). *Mind and Morals: Essays on Ethics and Cognitive Science,* Cambridge, MA: MIT Press.

Mayer, E. A. (2011). Gut feelings: the emerging biology of gut-brain communication. *Nature Reviews Neuroscience* 12, 453–66.

McCarthy, J. (1979). Ascribing mental qualities to machines. In: Ringle, M. (Ed.) *Philosophical Perspectives in Artificial Intelligence.* London: Humanities Press, pp. 161–95.

McCarthy, J. and Hayes, P. (1969). Some philosophical problems from the standpoint of artificial intelligence. In: Meltzer, B. and Michie, D. (Eds.) *Machine Intelligence,* Vol 4. Edinburgh: Edinburgh University Press, pp. 463–502.

McClelland, J. L. and Rumelhart, D. E. and the PDP Research Group (1986). *Parallel Distributed Processing: Explorations in the Microstructure of Cognition, Vol. 2: Psychological and Biological Models.* Cambridge, MA: MIT Press.

McCulloch, W. S. and Pitts, W. (1943). A logical calculus of the ideas immanent in nervous activity. *Bulletin of Mathematical Biophysics* 5, 115–33.

McDermott, D. (1987). We've been framed: Why AI is innocent of the frame problem. In: Pylyshyn, Z. (Ed.) *The Robot's Dilemma: The Frame Problem in Artificial Intelligence.* Norwood, NJ: Ablex, pp. 139–49.

Medick, V. (2013). 'Credible Deterrence': Germany plans to deploy armed drones. *Spiegel Online,* 25 January. http://www.spiegel.de/international/ germany/germany-plans-to-deploy-armed-drones-in-combat- abroad-a-879633.html (accessed 6 February 2013).

Melnikoff, D. E. and Bargh, J. A. (2018). The mythical number 2. *Trends in Cognitive Sciences* 22(4), 280–93.

Mercier, H. and Sperber, D. (2017). *The Enigma of Reason.* Cambridge, MA: Harvard University Press.

Meredith, C. A. (1953). Single axioms for the systems (c,n), (c, 0), and (a,n) of the two-valued propositional calculus. *Computing Systems* 1, 155–64.

Merleau-Ponty, M. (1945/2002). *Phenomenology of Perception*. Oxon, UK: Routledge.

Metzinger, T. (2003). *Being No One: The Self-Model Theory of Subjectivity*. Cambridge, MA: MIT Press.

Miller, J. D., Scott, E. C. and Okamoto, S. (2006). Public acceptance of evolution. *Science* 313, 765–6.

Minsky, M. (1974). A Framework for Representing Knowledge, MIT AI Laboratory Memo 306. Reprinted in Winston, P., Ed. (1975). *The Psychology of Computer Vision*. New York: McGraw-Hill pp. 170–95. Also available at: https://courses.media.mit.edu/2004spring/mas966/Minsky%201974%20Framework%20for%20knowledge.pdf (accessed 24 June 2019).

Minsky, M. L. and Papert, S. A. (1969). *Perceptrons*. Cambridge, MA: MIT Press.

Moor, J. H. (1985). What is computer ethics? *Metaphilosophy* 16(4), 266–75.

Moore, G. E. (1925). A defense of common sense. In: Muirhead, J. H. (Ed.) *Contemporary British Philosophy*. London: Allen and Unwin, pp. 191–224.

Moravčík, M., Schmid, M., Burch, N. V. et al. (2017). DeepStack: Expert-level artificial intelligence in heads-up no-limit poker, *Science* 356, 508–13. And see: https://www.deepstack.ai.

Morgan, W. (2019). Trump scraps requirement to report some air strikes, *Politico*, 6 March. https://www.politico.com/story/2019/03/06/trump-civilian-deaths-drone-strikes-1207409 (accessed 24 June 2019).

Murphy, G. L. (2002). *The Big Book of Concepts*. Cambridge, MA: MIT Press.

Nagel, T. (1974). What is it like to be a bat? *Philosophical Review* 83, 435–50. Reprinted in Nagel (1979a), pp. 165–80.

Nagel, T. (1979a). *Mortal Questions*. Cambridge, UK: Cambridge University Press.

Nagel, T. (1979b). Subjective and objective. In Nagel (1979a), pp. 196–213.

Nass, C. and Brave, S. (2006). *Wired for Speech*. Cambridge, MA: MIT Press.

Nersessian, N. (2008). *Creating Scientific Concepts*. Cambridge, MA: MIT Press.

Newell, A. (1980). Physical Symbol Systems. *Cognitive Science* 4(2), 135–83.

Newell, A. and Simon, H. (1976). Computer science as empirical inquiry: Symbols and search. *Communications of the ACM* 9(3), 113–26.

Nielsen, M. A. and Chuang, I. L. (2000). *Quantum Information and Quantum Computation*. Cambridge: Cambridge University Press.

Nilsson, N. (2007). The physical symbol system hypothesis: Status and prospects. In Lungarella, M., Iida, F., Bongard, J. and Pfeifer, R. (Eds.) *50 Years of Artificial Intelligence. Lecture Notes in Computer Science*, Vol. 4850. Berlin: Springer, pp. 9–17.

Nilsson, N. (2010). *The Quest for Artificial Intelligence: A History of Ideas and Achievements*. Cambridge: Cambridge University Press.

Noë, A. (2009). *Out of Our Heads: Why You Are Not Your Brain, and Other Lessons from the Biology of Consciousness*. New York: Hill and Wang.

Nolen-Hoeksema, S. (2000). The role of rumination in depressive disorders and mixed anxiety/depressive symptoms. *Journal of Abnormal Psychology* 109(3), 504–11.

Northoff, G., Heinzel, A., de Greck, R., Bernpohl, F., Dobrowolny, H. and Panksepp, J. (2006). Self-referential processing in our brain – A meta-analysis of imaging studies on the self. *NeuroImage* 31, 440–57.

O'Brien, W. (1996). Meaning and mattering, *Southern Journal of Philosophy* 34, 339–60.

Orlov, Y. F. (1982). The wave logic of consciousness: A hypothesis. *International Journal of Theoretical Physics* 21, 37–53.

Otterbacher, J. and Talias, M. (2017). S/He's too warm/agentic!: The influence of gender on uncanny reactions to robots. *Proceedings of the 2017 ACM/IEEE International conference on HRI*, pp. 214–23.

Oudeyer, P.-Y. and Kaplan, F. (2007). What is intrinsic motivation? A typology of computational approaches. *Frontiers in Neurorobotics* 1, 6.

Palombo, D. J., Sheldon, S. and Levine, B. (2018). Individual differences in autobiographical memory. *Trends in Cognitive Sciences* 22(7), 583–97.

Panksepp, J. (2005). Affective consciousness: Core emotional feelings in animals and humans. *Consciousness and Cognition* 14(1), 30–80.

Pautz, A. (2013). Does Phenomenology Ground Mental Content? In: Kriegel, U. (Ed.) *Phenomenal Intentionality*. Oxford: Oxford University Press, pp. 194–234.

Peirce, C. S. (1885). On the algebra of logic: A contribution to the philosophy of notation. *American Journal of Mathematics* 7, 197–202.

Penn, D. C., Holyoak, K. J. and Povinelli, D. J. (2008). Darwin's mistake: Explaining the discontinuity between human and nonhuman minds. *Behavioral and Brain Sciences* 31, 109–78.

Penrose, R. (1989). *The Emperor's New Mind: Concerning Computers, Minds, and the Laws of Physics*. Oxford: Oxford University Press.

Penrose, R. (1994). *Shadows of the Mind*. Oxford, UK: Oxford University Press.

Picard, R. W. (2010). Affective Computing: From laughter to IEEE. *IEEE Transactions of Affective Computing* 1(1), 11–17.

Picard, F. and Kurth, F. (2014). Ictal alterations of consciousness during ecstatic seizures. *Epilepsy & Behavior* 30, 58–61.

Plato (360 BCE). Sophist. http://www.perseus.tufts.edu/hopper/text?doc=Plat.%20Soph.

Port, R. F. and van Gelder, T., Eds. (1995). *Mind as Motion: Explorations in the Dynamics of Cognition*. Cambridge, MA: The MIT Press.

Putnam, H. (1960). Minds and Machines. In: Hook, S. (Ed.) *Dimensions of Mind: A symposium*. New York: New York University Press, pp. 57–70.

Putnam, H. (1992). Why functionalism didn't work. In: Earman, J. (Ed.) *Inference, Explanation, and Other Frustrations: Essays in the Philosophy of Science*. Berkeley, CA: University of California Press, pp. 255–70.

Qazi, S. H. and Jillani, S. (2012). Four myths about drone strikes. *The Diplomat*, 9 June. http://thediplomat.com/2012/06/09/four-myths-about-drone-strikes (accessed 28 March 2013).

Qin, P. and Northoff, G. (2011). How is our self related to midline regions and the default-mode network? *NeuroImage* 57, 1221–33.

Rapaport, W. J. (1988). Syntactic semantics: Foundations of computational natural-language understanding. In: Fetzer, J. H. (Ed.), *Aspects of Artificial Intelligence*, Dordrecht: Kluwer Academic Publishers, pp. 81–131.

Rapaport, W. J. (1995). Understanding understanding: Syntactic semantics and computational cognition. In: Tomberlin, J. E. (Ed.), *AI, Connectionism, and Philosophical Psychology, Philosophical Perspectives*, Vol. 9, Atascadero, CA: Ridgeview, pp. 49–88. Reprinted in Clark, A. and Toribio, J., Eds. (1998). *Language and Meaning in Cognitive Science: Cognitive Issues and Semantic Theory, Artificial Intelligence and Cognitive Science: Conceptual Issues*, Vol. 4. Hamden, CT: Garland, pp. 73–112.

Rapaport, W. J. (1999). Implementation is semantic interpretation. *The Monist* 82, 109–30. Longer version preprinted as *Technical Report 97-15* SUNY Buffalo Department of Computer Science and *Technical Report 97-5* SUNY Buffalo Center for Cognitive Science.

Rapaport, W. J. (2000). How to pass a Turing Test: Syntactic semantics, natural-language understanding, and first-person cognition. *Journal of Logic, Language, and Information* 9, 467–90.

Rapaport, W. J. (2006). How Helen Keller used syntactic semantics to escape from a Chinese room. *Minds & Machines* 16(4), 381–436.

Reik, L. and Howard, D. (2014). A code of ethics for the human-robot interaction profession. *Proceedings of We, Robot 2014*. Miami, FL: University of Miami School of Law.

Rochat, P. (2012). Primordial sense of embodied self-unity. In: Slaughter, V. and Brownell, C. A. (Eds.) *Early Development of Body Representations*. Cambridge: Cambridge University Press (pp 3–18).

Rogers, M. L. (2007). Action and inquiry in Dewey's philosophy. *Transactions of the Charles S. Pierce Society* 43(1), 90–115.

Rosenblatt, F. (1958). The perceptron: A probabilistic model for information storage and organization in the brain. *Psychological Review* 65, 386–408.

Rugg, M. D. and Viberg, K. L. (2013). Brain networks underlying episodic memory retrieval. *Current Opinion in Neurobiology* 23(2), 255–60.

Rumelhart, D. E., McClelland, J. L. and PDP Research Group (1986). *Parallel Distributed Processing: Explorations in the Microstructure of Cognition, Vol. 1: Foundations*. Cambridge, MA: MIT Press.

Ryle, G. (1949). *The Concept of Mind*. Chicago: The University of Chicago Press.

Samuel, A. L. (1963). Some studies in machine learning using the game of checkers. In: Feigenbaum, E. A. and Feldman, J. (Eds.) *Computers and Thought*. New York: McGraw-Hill, pp. 71–105.

Samuel, A. L. (1967). Some studies in machine learning using the game of checkers. II. Recent progress. *IBM Journal of Research and Development* 11, 601–17.

Sauer, F. and Schörnig, N. (2012). Killer drones – The silver bullet of democratic warfare? *Security Dialogue* 43(4), 363–80.

Schank, R. and Abelson, R. (1977). *Scripts, Plans, Goals, and Understanding: An Inquiry into Human Knowledge Structures*. Norwood, NJ: Lawrence Erlbaum Associates.

Scheutz, M. and Crowell, C. (2007). The Burden of Embodied Autonomy: Some Reflections on the Social and Ethical Implications of Autonomous Robots. *ICRA 2007 Workshop on Roboethics*. Rome: IEEE.

Schilbach, L., Eickhoff, S. B., Rotarska-Jagiela, A., Fink, G. R. & Vogeley, K. (2008). Minds at rest? Social cognition as the default mode of cognizing and its putative relationship to the 'default system' of the brain. *Consciousness and Cognition* 17, 457–67.

Scholl, B. J. (2007). Object persistence in philosophy and psychology. *Mind & Language* 22, 563–91.

Schwabe, L., Nader, K. and Pruessner, J. C. (2014). Reconsolidation of human memory: Brain mechanisms and clinical relevance. *Biological Psychiatry* 76, 274–80.

Schwartz, J. M., Stapp, H. P. and Beauregard, M. (2005). Quantum physics in neuroscience and psychology: A neurophysical model of mind–brain interaction. *Philosophical Transactions of the Royal Society B* 360, 1309–27.

Searle, J. R. (1980). Minds, brains and programs. *Behavioral and Brain Sciences* 3(3), 417–24.

Searle, J. (1990). Is the brain a digital computer? *Proceedings and Addresses of the American Philosophical Association* 64: 21–37.

Searle, J. (1992). *The Rediscovery of the Mind*. Cambridge, MA: MIT Press.

Sellars, W. (1962). Philosophy and the scientific image of man. In: Colodny, R. (Ed.) *Frontiers of Science and Philosophy*. Pittsburgh, PA: University of Pittsburgh Press, pp. 35–78.

Sergent, C., Baillet, S. and Dehaene, S. (2005). Timing of the brain events underlying access to consciousness during the attentional blink. *Nature Neuroscience* 8, 1391–400.

Shanahan, M. (1997). *Solving the Frame Problem: A Mathematical Investigation of the Common Sense Law of Inertia*. Cambridge, MA: MIT Press.

Shanahan, M. (2012). The brain's connective core and its role in animal cognition. *Philosophical Transactions of the Royal Society B* 367, 2704–14.

Shanahan, M. (2016). The Frame Problem. In: Zalta, E. N. (Ed.) *The Stanford Encyclopedia of Philosophy* (Spring 2016 Edition), https://plato.stanford.edu/archives/spr2016/entries/frame-problem (accessed 24 June 2019).

Shannon, C. E. (1948). A Mathematical Theory of Communication. *The Bell System Technical Journal* 27, 379–423, 623–56.

Shapiro, L. (2011). *Embodied Cognition*, New York: Routledge.

Sharkey, N. (2008). Grounds for discrimination: Autonomous robot weapons. *RUSI Defence Systems* 11(2), 86–9.

Sharkey, N. (2009). Death strikes from the sky. *IEEE Technology and Society Magazine* 28(1): 16–19.

Sharkey, N. (2010). Saying 'no!' to lethal autonomous targeting. *Journal of Military Ethics* 9(4): 369–83.

Sharkey, N. (2011). Moral and legal aspects of military robots. In: Dabringer, G. (Ed.) *Ethical and Legal Aspects of Unmanned Systems*. Vienna: Institut für Religion und Frieden, pp. 43–51.

Sharkey, N. (2012). Killing made easy: From joysticks to politics. In: Lin, P., Abney, K. and Bekey, G. A. (Eds.) *Robot Ethics: The Ethical and Social Implications of Robotics*. Cambridge, MA: MIT Press, pp. 111–28.

Sharkey, N. (2018). Mama Mia, It's Sophia: A show robot or dangerous platform to mislead? *Forbes*, 17 November. https://www.forbes.com/sites/noelsharkey/2018/11/17/mama-mia-its-sophia-a-show-robot-or-dangerous-platform-to-mislead (accessed 24 June 2019).

Shaw-Garlock, G. (2014). Gendered by design: Gender codes in social robotics. *Frontiers in Artificial Intelligence and Applications* 273, 309–17.

Sheline, Y. I., Barch, D. M., Price, J. L., Rundle, M. M., Vaishnavi, S. N., Snyder, A. Z., Mintun, M. A., Wang, S., Coalson, R. S. and Raichle, M. E. (2009). The default mode network and self-referential processes in depression. *Proceeding of the National Academy of Sciences USA* 106(6), 1942–7.

Shettleworth, S. J. (2010). *Cognition, Evolution and Behavior.* (2nd Ed.) Oxford University Press, New York.

Shor, P. W. (1994). Algorithms for quantum computation: Discrete logarithms and factoring. *Proceedings of the 35th Annual Symposium on Foundations of Computer Science,* Santa Fe, NM (IEEE) pp. 124–34.

Siegel, M., Brazeal, C. and Norton, M. I. (2009). Persuasive Robotics: The Influence of Robot Gender on Human Behavior. *2009 IEEE/RSJ International Conference on Intelligent Robots and Systems*, pp. 2563–8.

Siewert, C. (2017). Consciousness and Intentionality. In: Zalta, E.N. (Ed.) *The Stanford Encyclopedia of Philosophy* (Spring 2017 Edition), https://plato.stanford.edu/archives/spr2017/entries/consciousness-intentionality (accessed 25 June 2019).

Silver, D., Huang, A., Maddison, C. J. et al. (2016). Mastering the game of Go with deep neural networks and tree search. *Nature* 529, 484–9.

Simon, J., Ed. (2020). *The Routledge Handbook of Trust and Philosophy*. New York and London: Routledge

Simons, J. S., Henson, R. N. A., Gilbert, S. J. and Fletcher, P. C. (2008). Separable forms of reality monitoring supported by anterior prefrontal cortex. *Journal of Cognitive Neuroscience* 20(3), 447–57.

Singer, P. W. (2009). *Wired For War*. New York: Penguin Press.

Sipser, M. (2013). *Introduction to the Theory of Computation, Third Edition*. Boston: MA: Cengage Learning.

Sloman, A. (1978). *The Computer Revolution in Philosophy: Philosophy, Science, and Models of the Mind*. Hemel Hempstead: Harvester Press.

Sloman, A. (1994). Semantics in an intelligent control system. *Philosophical Transactions of the Royal Society A* 349, 43–58.

Smith, J. E. and Nair, R. (2005). The architecture of virtual machines. *IEEE Computer* 38(5), 32–8.

Smolensky, P. (1988). On the proper treatment of connectionism. *Behavioral and Brain Sciences* 11, 1–23.

Sousa, A. et al. (2017). Molecular and cellular reorganization of neural circuits in the human lineage. *Science* 358, 1027–32.

Sparrow, R. W. (2007). Killer robots. *Journal of Applied Philosophy* 24(1), 62–77.

Sparrow, R. W. (2009a). Predators or plowshares? Arms control of robotic weapons. *IEEE Technology and Society Magazine* 28(1): 25–9.

Sparrow, R. W. (2009b). Building a better WarBot: Ethical issues in the design of unmanned systems for military applications. *Science and Engineering Ethics* 15, 169–87.

Sparrow, R. W. (2011). The ethical challenges of military robots. In: Dabringer, G. (Ed.) *Ethical and Legal Aspects of Unmanned Systems*. Vienna: Institut für Religion und Frieden, 87–102.

Spivey, M. (2007). *The Continuity of Mind*. Oxford: Oxford University Press.

Sporns, O. and Honey, C. J. (2006). Small worlds inside big brains. *Proceedings of the National Academy of Sciences USA* 111, 15220–5.

Suddendorf, T. and Corballis, M. T. (2007). The evolution of foresight: What is mental time travel, and is it unique to humans? *Behavioral and Brain Sciences* 30(3), 299–313.

Sullins, J. P. (2010a). RoboWarfare: Can robots be more ethical than Humans on the battlefield? *Ethics and Information Technology* 12(3), 263–75.

Sullins, J. P. (2010b). Rights and computer ethics. In: Floridi, L. (Ed.) *The Cambridge Handbook of Information and Computer Ethics*. Cambridge: Cambridge University Press, pp. 116–33.

Sullins, J. P. (2011). Aspects of telerobotic systems. In: Dabringer, G. (Ed.) *Ethical and Legal Aspects of Unmanned Systems*. Vienna: Institut für Religion und Frieden, pp. 157–67.

Sullins, J. (2015). Applied ethics for the reluctant roboticist. *Proceedings of the Emerging Policy and Ethics of Human-Robot Interaction Workshop at HRI*, Portland, OR.

Sullins, J. (2020). Trust in Robots. In Simon, J. (Ed.) *The Routledge Handbook of Trust and Philosophy*. New York and London: Routledge

Sun, R. (2007). The importance of cognitive architectures: An analysis based on CLARION. *Journal of Experimental and Theoretical Artificial Intelligence* 19(2), 159–93.

Sun, R. (2013). Autonomous generation of symbolic representations through subsymbolic activities. *Philosophical Psychology* 26(6), 888–912.

Sun, R. (2018). Why is a computational framework for motivational and metacognitive control needed? *Journal of Experimental and Theoretical Artificial Intelligence* 30(1), 13–37.

Sun, R. and Zhang, X. (2006). Accounting for a variety of reasoning data within a cognitive architecture. *Journal of Experimental and Theoretical Artificial Intelligence* 18(2), 169–91.

Sun, R., Slusarz, P. and Terry, C. (2005). The interaction of the explicit and the implicit in skill learning: A dual-process approach. *Psychological Review* 112(1), 159–92.

Taddeo, M. (2010). Modelling trust in artificial agents, a first step toward the analysis of E-Trust. *Minds & Machines* 20(2), 243–57.

Taddeo, M. and Floridi, L. (2005). Solving the symbol grounding problem: A critical review of fifteen years of research. *Journal of Experimental and Theoretical Artificial Intelligence* 17, 419–45.

Taddeo, M. and Floridi, L. (2011). The case for E-trust. *Ethics and Information Technology* 13, 1–3.

Tang, Y.-Y., Hölzel, B. K. and Posner, M. I. (2015). The neuroscience of mindfulness meditation. *Nature Reviews Neuroscience* 16, 213–25.

Thayer, J. F. and Lane, R. D. (2009). Claude Bernard and the heart–brain connection: Further elaboration of a model of neurovisceral integration. *Neuroscience & Biobehavioral Reviews* 33(2), 81–8.

Thelen, E. and Smith, L. B. (1994). *A Dynamic Systems Approach to the Development of Cognition and Action*. Cambridge, MA: MIT Press.

Thelen, E., Schöner, G., Scheier, C. and Smith, L. (2001). The dynamics of embodiment: A field theory of infant perseverative reaching. *Behavioral and Brain Sciences* 24, 1–86.

Thompson, E. (2001). Empathy and consciousness. *Journal of Consciousness Studies* 8(5), 1–32.

Thompson, E. (2007). *Mind in Life: Biology, Phenomenology, and the Sciences of the Mind*, Cambridge, MA: Harvard University Press.

Tononi, G. (2008). Consciousness as integrated information: A provisional manifesto. *Biological Bulletin* 215, 216–42.

Tononi, G. and Koch, C. (2015). Consciousness: Here, there and everywhere? *Philosophical Transactions of the Royal Society B* 370, 20140167.

Torrance, S. (2008). Ethics and consciousness in artificial agents. *AI & Society* 22, 495–521.

Torrance, S. (2011). Machine ethics and the idea of a more-than-human moral world. In: Anderson, M. and Anderson, S. (Eds.) *Machine Ethics*. Cambridge: Cambridge University Press, pp. 115–37.

Torrance, S. (2014). Artificial consciousness and artificial ethics: Between realism and social relationism. *Philosophy of Technology* 27, 9–29.

Trevarthen, C. (2010). What is it like to be a person who knows nothing? Defining the active intersubjective mind of a newborn human being. *Infant and Child Development* 20(1), 119–35.

Turing, A. M. (1936). On computable numbers, with an application to the *Entscheidungsproblem*. *Proceedings of the London Mathematical Society* 2–42, 230–65.

Turing, A. M. (1950). Computing machinery and intelligence. *Mind* 59, 433–60.

Vallor, S. (2016). *Technology and the Virtues: A Philosophical Guide to a Future Worth Wanting*. Oxford: Oxford University Press.

van Gelder, T. (1995). What might cognition be, if not computation? *Journal of Philosophy* 92(7), 345–81.

van Gelder, T. (1998). The dynamical hypothesis in cognitive science. *Behavioral and Brain Sciences* 21(5), 637–8.

van Heijenoort, J. (1967). *From Frege to Gödel: A Source Book in Mathematical Logic, 1879–1931*. Cambridge, MA: Harvard University Press.

Varela, F., Thompson, E. and Rosch, E. (1991). *The Embodied Mind: Cognitive Science and Human Experience*. Cambridge, MA: MIT Press.

Viscelli, S. (2018). Driverless? Autonomous Trucks and the Future of the American Trucker. UC Berkeley Center for Labor Research and Education and Working Partnerships USA, September. http://driverlessreport.org/files/driverless.pdf (accessed 25 June 2019).

von Neumann, J. (1958). *Mathematical Foundations of Quantum Mechanics*. Princeton, NJ: Princeton University Press.

von Uexküll, J. (1957). A stroll through the worlds of animals and men. In: Schiller, C. (Ed.), *Instinctive Behavior*. New York: International Universities Press, pp. 5–80.

Wallach, W. and Allen, C. (2009). *Moral Machines: Teaching Robots Right From Wrong*. New York: Oxford University Press.

Ward, J. (2013). Synesthesia. *Annual Review of Psychology* 64, 49–75.

Watson, J. B. (1913). Psychology as the behaviorist views it. *Psychological Review* 20, 158–77.

Weizenbaum, J. (1976). *Computer Power and Human Reason: From Judgment To Calculation*, San Francisco: W. H. Freeman.

Wheeler, J. A. (1983). Law without law. In: Wheeler, J. A. and Zurek, W. H. (Eds.) *Quantum Theory and Measurement*. Princeton, NJ: Princeton University Press, pp. 182–213.

White, N. P., Trans. (1993). *Plato: Sophist*. Indianapolis, IN: Hackett.

Whitehead, A. N. and Russell, B. (1910). *Principia Mathematica, Vol. 1*. Cambridge: Cambridge University Press.

Whitehead, A. N. and Russell, B. (1912). *Principia Mathematica, Vol. 2*. Cambridge: Cambridge University Press.

Wigner, E. P. (1961). Remarks on the mind-body question. In Good, I. (Ed.) *The Scientist Speculates*. London: Heineman, pp. 284–302.

Wilson, A. and Golonka, S. (2013). Embodied cognition is not what you think it is. *Frontiers in Psychology* 4, 1–13.

Wilson, M. (2002). Six views of embodied cognition. *Psychonomic Bulletin and Review* 9(4), 625–36.

Winner, L. (1978). *Autonomous Technology: Technics-Out-of-Control As a Theme in Political Thought*. Cambridge, MA: MIT Press.

Wittgenstein, L. (1953). *Philosophical Investigations*. New York: Macmillan.

Wootson, C. R., Jr. (2017). Saudi Arabia, which denies women equal rights, makes a robot a citizen. *The Washington Post*, 29 October. https://www.washingtonpost.com/news/innovations/wp/2017/10/29/saudi-arabia-which-denies-women-equal-rights-makes-a-robot-a-citizen (accessed 25 June 2019).

Ying, M. (2010). Quantum computation, quantum theory and AI. *Artificial Intelligence* 174, 162–76.

Zubair Shah, P. (2012). My drone war. *Foreign Policy*, March/April. http://www.foreignpolicy.com/articles/2012/02/27/my_drone_war?page=0,4 (accessed 1 February 2013).

Index